# The identification of flowering plant families

SECOND EDITION

# The identification of flowering plant families

*including a key to those native and cultivated in north temperate regions*

P. H. DAVIS D.SC.
READER IN BOTANY, UNIVERSITY OF EDINBURGH

J. CULLEN PH.D.
ASSISTANT KEEPER, ROYAL BOTANIC GARDEN, EDINBURGH

SECOND EDITION

CAMBRIDGE UNIVERSITY PRESS
CAMBRIDGE
LONDON · NEW YORK · MELBOURNE

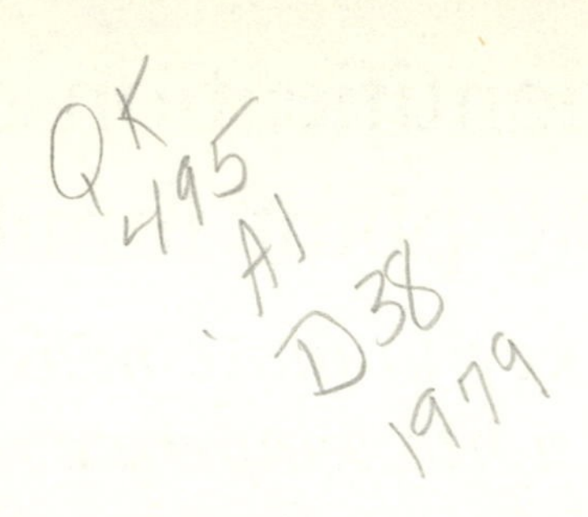


Published by the Syndics of the Cambridge University Press
The Pitt Building, Trumpington Street, Cambridge CB2 1RP
Bentley House, 200 Euston Road, London NW1 2DB
32 East 57th Street, New York, NY 10022, USA
296 Beaconsfield Parade, Middle Park, Melbourne 3206, Australia



First published by Oliver & Boyd 1965

Second edition published by the Cambridge University Press 1979

Printed in Great Britain by
Cox & Wyman Ltd, London, Fakenham and Reading

*Library of Congress Cataloguing in Publication Data*

Davis, Peter Hadland.
The identification of flowering plant families, including a key to those native and cultivated in north temperate regions.

Includes index
1. Angiosperms – Identification. 2. Flowers – Identification.
I. Cullen, James, joint author. II. Title.
QK495.A1D38 1979 582'.13'09123 78–8125
ISBN 0 521 22111 0 hard covers
ISBN 0 521 29359 6 paperback

# CONTENTS

# ILLUSTRATIONS

# PREFACE

Since the first edition of this book was published (by Oliver & Boyd, Edinburgh, 1965), many changes have taken place in Flowering Plant classification. New information, mainly phytochemical and micromorphological, has been brought forward and applied to classification on an ever-increasing scale. New systems of classification have been developed to synthesise these results and to integrate them with the more traditional schemes. All these activities have rendered the first edition somewhat out of date, and, as the book has been out of print for the last seven years, we were delighted when the Cambridge University Press agreed to publish a second edition.

The need for a revised edition of the book has been made clear to us not only by the changes mentioned above, but also by the reduction of emphasis on the teaching of comparative morphology and taxonomy in schools and universities, owing to the increasing need for training in experimental disciplines. In addition, the book has evidently had a wider use than we expected. We have heard, for instance, of its use by students in Australia and New Zealand, and even by collectors in the highlands of Colombia.

We have taken the opportunity provided by a second edition to revise the whole of the text, taking into account the comments of reviewers, students and our colleagues. The introductory chapters have been revised and enlarged, particularly the illustrated section on Usage of Terms. These chapters provide basic information on plant morphology and how best to examine plant material for the purpose of identification. The keys themselves (considerably enlarged) are now presented in the more compact bracket form, as students seem to find this type of presentation easier to use than the indented type adopted in the first edition. As far as we are aware, this is still the only book available in which all this information, including a clearly presented key for family identification, is presented in such a self-contained form.

The taxonomic system employed in the arrangement and description of families has also been changed, as the Benson system, used in the first edition, does not take account of modern opinions. We have adopted the system put forward by G. L. Stebbins for two reasons: it seems to us to represent a reasonably widely held consensus of current thinking about Flowering Plant classification; and because of the imminent publication of *Flowering Plants of the World* (edited by V. H. Heywood, Oxford University Press, 1978), which uses the same system. Heywood's book, which contains descriptions of all the families, with numerous illustrations and maps, but no family keys, will provide a valuable complement to the present volume. The adoption of Stebbins's system (itself mainly a synthesis of those of Cronquist and Takhtajan), has necessitated certain rearrangements of the families recognised, and we have added some others (mainly exotic families sometimes seen in cultivation), so that the total number of families now included is 272.

We hope that this new edition will be useful to anyone who wants to name plants, whether he is a professional biologist, a keen gardener or an interested amateur. The text has been revised with students of botany in mind. The presentation of the information in key form, necessitating a series of choices between alternative morphological states, forces the student to make a close examination of plants. This helps to develop his observational powers, and assists his memory in retaining important characters. With family characters taught less and less, and the number of recognised families tending to increase, identification by a key rather than by remembering family characters parrot-fashion, becomes more necessary than ever. Although this book contains more families than most students are likely to meet, the discipline of running down a plant in a long key should result in the student observing more characters of the plant than he may notice when using a short one.

In our view a natural classification of plants and their correct identification (for which the families are an important step on the road) remain essential for the progress of biology on a broad front.

Edinburgh
December 1977

# ACKNOWLEDGEMENTS

In the preparation of this book we have been assisted by many botanists, and we gratefully acknowledge their help: Professor J. P. M. Brenan, B. L. Burtt, Professor C. D. K. Cook, Dr T. R. Dudley, Professor V. H. Heywood, R. D. Meikle, Professor H. Merxmüller, Professor T. G. Tutin, Dr S. M. Walters and Dr P. F. Yeo. We are also indebted to our students and others, particularly Miss M. M. MacDonald, who, by attempting to run down numerous plants in the key, enabled us to improve on the first edition. To D. M. Henderson, Regius Keeper of the Royal Botanic Garden, Edinburgh, we are grateful for generous facilities. We are especially indebted to M. J. E. Coode who drew the illustrations for the first edition and gave us much help in other ways; and to Miss R. M. Smith, who drew the illustrations for this edition.

# ABBREVIATIONS

Mostly used in the short descriptions of families on pp. 64–96

| | | | |
|---|---|---|---|
| A | androecium | *N* | North |
| alt | alternate | n | many |
| antipet | antipetalous | opp | opposite |
| ax | axile placentation | ov | ovules |
| bisex | bisexual | P | perianth |
| C | corolla | par | parietal placentation |
| *C* | Central | perig | perigynous |
| *E* | East | rar | rarely |
| epipet | epipetalous | *S* | South |
| epig | epigynous | s.l. | *sensu lato* |
| excl | excluding | solit | solitary |
| exstip | exstipulate | s.s. | *sensu stricto* |
| fl | flower | stip | stipules, stipulate |
| fr | fruit | sup | superior |
| G | gynoecium | Temp | temperate |
| hypog | hypogynous | Trop | tropical |
| incl | including | unisex | unisexual |
| inf | inferior | usu | usually |
| infl | inflorescence | var | various |
| Is | Islands | *W* | West |
| K | calyx | zyg | zygomorphic |
| loc | locular | / | or |
| lvs | leaves | ± | more or less |

# INTRODUCTION

The identification of the family to which a plant belongs is usually the first essential step in its complete identification. This key attempts to provide a means of identification for all Angiosperm families native or cultivated, out-of-doors or under glass, in north temperate regions. In practice, we have taken the southern limit of our area as approximately 30° N, thus excluding all of Mexico and Florida in the New World and most of India and subtropical China in the Old World. A few, mainly tropical families with a small number of genera native in China north of 30° N have also been excluded as, unless cultivated in Europe or North America, they are so infrequently seen. The key covers 272 families.

As far as cultivated plants are concerned, a few tropical families which, although sometimes grown, rarely flower in cultivation (e.g. *Dipterocarpaceae*), have been excluded. The key has been constructed to allow for the identification of the frequently cultivated representatives of tropical or southern hemisphere families; it may not work for other genera. No attempt has been made to cope with the double-flowered, wilder excesses of the plant breeder. We should be glad to hear from readers of any familiar genus that does not 'key out' to its family.

The taxonomic system followed in the book is that given by G. L. Stebbins (*Flowering Plants: Evolution above the Species Level*, 1974), though with a few modifications (given in detail on p. 63). G. H. M. Lawrence's *Taxonomy of Vascular Plants* (1951), J. Hutchinson's *The Families of Flowering Plants*, 2nd edn (1959), the 12th edition (ed. H. Melchior) of Engler's *Syllabus der Pflanzenfamilien* (1964) and H. K. Airy Shaw's editions (7th edn, 1966 and 8th edn, 1973) of Willis's *Dictionary of Flowering Plants and Ferns*, have all been frequently consulted in the preparation of the key and descriptions; in addition the systems presented by A. Cronquist (*The Evolution and Classification of Flowering Plants,* 1968), A. L. Takhtajan *(Flowering Plants: Origin and Dispersal*, English edition, translated by C. Jeffrey, 1969) and R. Thorne (A phylogenetic classification of the Angiospermae,

*Evolutionary Biology* **9** (1976) 35–106) have been valuable in deciding which genera belong to each family. Three families – *Leguminosae, Saxifragaceae* and *Liliaceae* – are keyed out in a very broad sense in the main key; we have given separate keys to their segregates (about which there is often dispute as to status).

The short descriptions of the families are intended both as a check on identification and as a terse presentation of the important family characters. These descriptions refer to the whole family, not only to those representatives catered for in the key. The distribution of each family has been given, though necessarily without any great detail. The arrangement of the families should serve as a rough guide to their relationships, although it must be realised that these may cut across the limits of the orders and superorders, whose circumscription is largely a matter of individual opinion.

To make the book as useful and self-contained as possible we have provided a glossary; terms requiring a more extensive explanation are discussed in the section entitled 'Usage of Terms' (p. 4). For readers not particularly experienced in plant identification, we have included a section on 'Examining the Plant'. This details many of the important characters easily overlooked by a too superficial examination of specimens, and should help to avoid 'howlers' in identification.

The short section on 'Further Identification' is intended to help the reader to obtain the generic and specific names of his plants. For reasons of space this section is short, and refers only to major works and lists.

The student should not assume that, by keying out one genus, he will necessarily become acquainted with the characters of the family to which it belongs; to know *Prunus*, for instance, is to know very little about the *Rosaceae*! It is not possible to make a workable key to the families in which each family keys out only once; in this key many families key out several times, and a reference to each appearance of a family in the key is given in the index. The index also includes a number of synonyms, as well as the names of some families, often recognised as separate entities, which we have included in larger families.

It is not possible in a book of this size to give information about the theory and practice of classification. It must be remembered that the name of a plant is only a key to further information about it, not an end in itself. Any reader whose interest in this aspect of the subject is stimulated by the practice of identifying plants, will find a comprehensive survey in Davis & Heywood's *Principles of Angiosperm Taxonomy*

(1963, reprinted by Krieger, New York, 1973), and R. H. Crowson's *Classification and Biology* (1970). More recent theoretical developments are discussed in Sneath & Sokal's *Numerical Taxonomy* (1973), and in the pages of such journals as *Evolution*, *Systematic Botany*, *Systematic Zoology*, *Taxon*, etc.

# USAGE OF TERMS

A few terms need a fuller explanation than can be given in the Glossary; these are the terms referring to the relative positions of the floral organs, placentation and aestivation, which are discussed and illustrated below.

It must be emphasised that in this section we have not so much attempted to *define* terms as to explain how they are used in this book. This has been necessary because there is little consistency in their usage in the literature. We have avoided, as much as possible, explanations which require the acceptance of particular theories as to the origins, homologies, etc., of structures, and have contented ourselves with a purely descriptive terminology which should be practicable and widely applicable whatever interpretation is put on the nature of the floral parts.

## HYPOGYNY, PERIGYNY AND EPIGYNY

Two sets of terms are used in describing the relative positions of the floral organs. One (superior/inferior) is generally used with reference to the position of the ovary with respect to the other floral organs; the other (hypogyny/perigyny/epigyny) appears to refer to the position of the other floral organs with respect to the ovary, but is perhaps best expressed in terms of the apparent adnation of parts, rather than with reference to the ovary position alone. The phenomena covered by both sets of terms are best seen in longitudinal sections of flowers (cf. figs 1–4).

Egler (*Chronica Botanica* **12** (1951) 169–73) has discussed both these sets of terms and has recommended their replacement. We have not followed his recommendations because our aim has been to refine and clarify the use of the standard terminology (which is used in different senses by different authors) rather than to try to replace it.

The terms referring to ovary position are not especially ambiguous – a *superior ovary* is one borne on the torus above the insertion of the other floral organs (regardless of whether these are free or united – cf. figs. 2 and 3 (*a*)–(*c*); an *inferior ovary* is one borne below the point of insertion of the other floral organs (free or united), so that they appear to be borne on or near the top of the ovary (cf. figs. 3(*d*), (*e*); 4(*a*), (*b*)). The only difficulty is introduced by the rather frequent occurrence of an intermediate condition in which the ovary is said to be *partly inferior*. In this condition the tops of the loculi of the ovary occur above the point of insertion of the other floral organs, and the bottoms of the loculi occur below this point (cf. fig. 3(*d*)).

The other terminology is more difficult to apply. One of the main difficulties with these terms is that they have frequently been applied to the flower as a whole. We have found this practice misleading (cf. table 1), and prefer to use the terms with reference to the perianth and stamens. We hope that our usage, which is based on that formulated by De Candolle in his *Théorie Elémentaire de Botanique* (1813), will be accepted as giving a less ambiguous description of the relationships of the floral parts.

Table 1 and the accompanying diagram (fig. 1) should make clear the usage of the terms that we have adopted, and show the relationship between the two sets of terms discussed above.

A few words of explanation may help to make the table easier to understand. The 'ring or collar of tissue' mentioned in the table, probably best termed a perigynous or epigynous zone, has often been referred to as a 'floral cup or tube' or 'hypanthium'. These terms have generally been used when sepals, petals and stamens are all inserted on such a ring of tissue, as in *Prunus* or *Rhamnus*.

In flowers with a superior ovary the stamens may be apparently borne on the corolla (e.g. *Primulaceae*, *Malvaceae*, etc.); in these cases it seems reasonable to describe the corolla plus androecium as perigynous and the calyx as hypogynous. In such flowers there is no term in general use for the tissue between the insertion of the corolla plus androecium on the torus and the insertion of the androecium on the corolla; however, there seems no reason why it should not be termed (if a term is needed) a *perigynous zone* (it is not, of course, possible to consider it a hypanthium). In other flowers the calyx and corolla are apparently adnate below, whereas the stamens are quite free (e.g. *Tropaeolum*). Again, there is no reason why the portion of tissue apparently composed of calyx and corolla should not be called a perigynous zone.

When the ovary is inferior, the stamens may be free from the corolla (e.g. *Vaccinium*) or adnate to it (e.g. *Compositae*). In the latter case, the tissue composed of corolla plus stamens may be termed an *epigynous*

Table 1. *Relationships of floral parts (cf.* fig. 1)

| Ovary (G) position | Fig. 1 | Insertion of perianth (P or K & C) and androecium (A) | Description adopted here | Description used in older literature |
|---|---|---|---|---|
| | (*a*) | PA or KCA inserted independently on the torus (e.g. *Ranunculus*) | PA or KCA hypogynous | Flower hypogynous |
| | (*b*) | K & C apparently adnate at the base, A inserted independently on the torus (e.g. *Tropaeolum*) | K & C perigynous, borne on a perigynous zone, A hypogynous | Various |
| G superior | (*c*) | C & A apparently adnate at the base, K inserted independently on torus (e.g. *Primula*) | K hypogynous, C & A perigynous, borne on a perigynous zone | Flower hypogynous, A epipetalous |
| | (*d*) | K, C & A inserted on a ring or collar of tissue which is inserted on the torus (e.g. *Prunus*) | K, C & A perigynous, borne on a perigynous zone | Flower perigynous |
| | (*e*) | P & A apparently adnate, C absent (e.g. *Daphne*) | P & A perigynous | Various |
| G partly inferior | (*f*) | P & A or K, C & A inserted independently, apparently on walls of ovary (e.g. *Paliurus*, some species of *Saxifraga*) | P & A or K, C & A partly epigynous | Various |
| | (*g*) | P & A or K, C & A inserted independently apparently on top of the ovary (e.g. *Umbelliferae*) | P & A or K, C & A epigynous | Flower epigynous |
| G fully inferior | (*h*) | K, C & A inserted on top of ovary, C & A adnate (e.g. *Viburnum*) | K, C & A epigynous, C & A borne on an epigynous zone | Flower epigynous, A epipetalous |
| | (*i*) | K, C & A inserted on a ring or collar of tissue itself inserted on top of the ovary (e.g. *Fuchsia*) | K, C & A epigynous, borne on an epigynous zone | Flower epigynous |

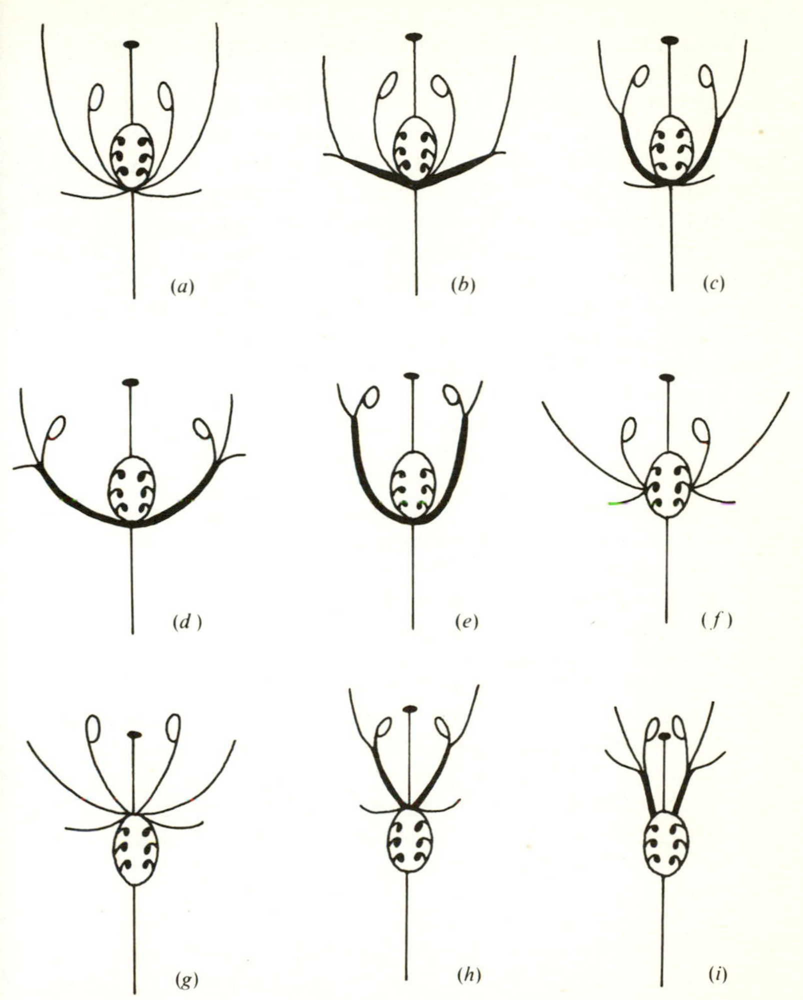


Figure 1. Diagram illustrating the usage of the terms hypogyny, perigyny and epigyny. Perigynous and epigynous zones are indicated by the use of heavy lines. For further information see table 1

*zone* (again, not equivalent to a hypanthium). In *Fuchsia* (*Onagraceae*) petals, sepals and stamens are all apparently borne on the top of such a zone.

A warning must be given here about unisexual flowers. If a pistillode is present in a male flower, it is possible to decide whether the whorls are hypogynous, perigynous or epigynous. If, however, no pistillode is present, care must be taken. When the ovary of the female flower is partly or completely inferior, it seems reasonable to treat the perianth and stamens of the male flower as epigynous; when the ovary in the female flower is superior, then, if the perianth and stamens in the male flower are inserted independently on the torus, they may be treated as hypogynous; if there is any apparent adnation, as perigynous. This is the procedure we have adopted in making the key. If the female flower lacks a perianth (e.g. *Betula*), then, of course, it is not possible to use these terms at all; we have used the term 'ovary naked' to refer to this condition.

The following situations occur in the families covered by the key:[1]

I. Perianth and stamens hypogynous; ovary superior (e.g. *Ranunculus*, *Geranium*, *Cistus*, *Silene*, etc.). Cf. fig. 2(*a*), (*b*).

II. Sepals hypogynous; petals and stamens perigynous; ovary superior (e.g. most *Malvaceae*, *Primulaceae*, *Scrophulariaceae*, etc.). Cf. fig. 2(*c*), (*d*).

III. Perianth perigynous; stamens hypogynous; ovary superior (e.g. *Tropaeolum*, *Aesculus*). Cf. fig. 2(*e*).

IV. Perianth and stamens perigynous; ovary superior (e.g. *Prunus*, *Geum*, *Bergenia*, *Staphylea*, *Daphne*, etc.). Cf. fig. 3(*a*)–(*c*).

V. Ovary partly or fully inferior; perianth and stamens epigynous, without an epigynous zone (e.g. *Umbelliferae*, *Campanula*, *Vaccinium*, etc.). Cf. fig. 3(*d*), (*e*).

VI. Ovary partly or fully inferior; perianth and stamens with an epigynous zone composed of 3 whorls (e.g. *Ribes*, *Fuchsia*, *Leptospermum*) or of 2 whorls (e.g. *Viburnum*, *Compositae*). Cf. fig. 4(*a*), (*b*).

As far as we know, perigynous and epigynous zones composed of all 3 whorls do not occur in gamopetalous flowers. In type V the stamens are sometimes adnate to the style (e.g. *Aristolochia*); all parts, however, are still epigynous. Types III and IV are often complicated in polypetalous families by the presence of an apparently *nectariferous disc* surrounding and sometimes almost covering the ovary. Usually, when such a disc is present, the perianth (as a clearly recognisable structure) is inserted on

[1] Other patterns occur in some tropical and southern hemisphere families.

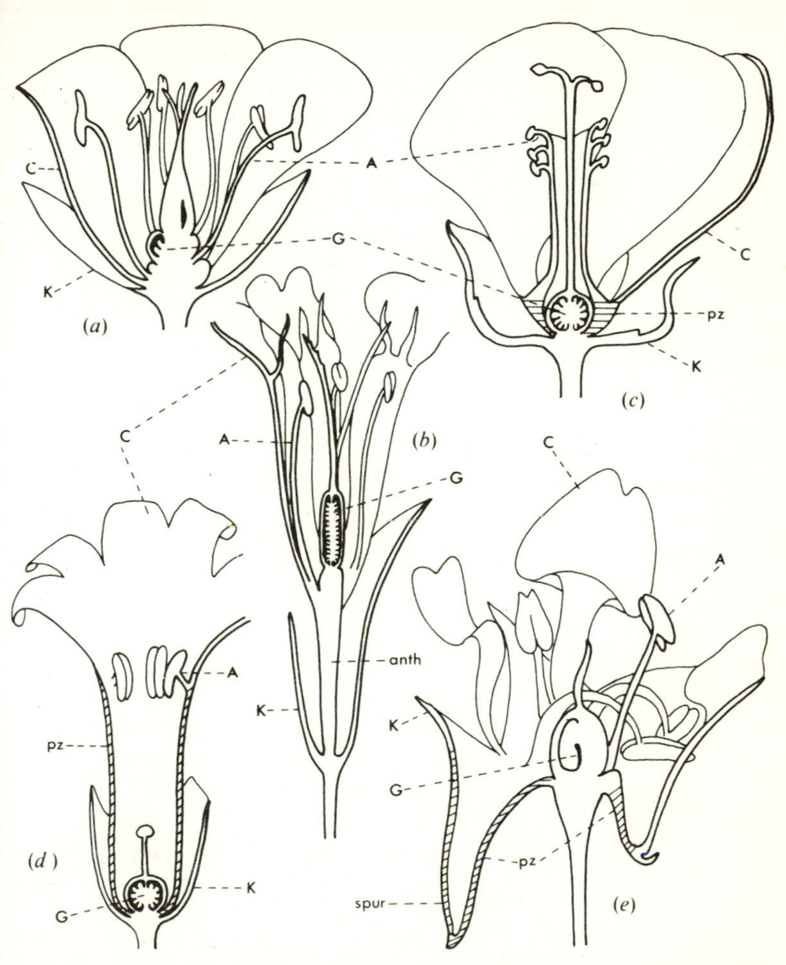


Figure 2. Relative positions of floral parts (cf. p. 8). Type I: (*a*) *Geranium*, (*b*) *Silene*; type II: (*c*) *Abutilon*, (*d*) *Primula*; type III: (*e*) *Tropaeolum*. A – androecium, anth – anthophore, C – corolla, G – gynoecium, K – calyx, pz – perigynous zone (shaded).

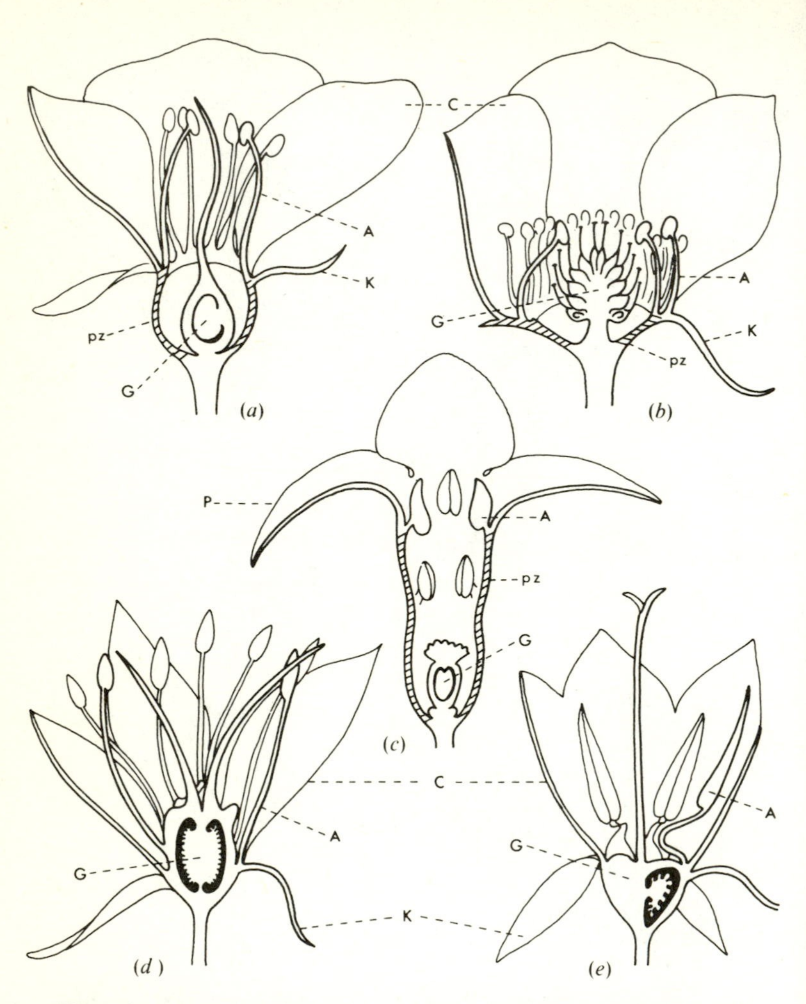


Figure 3. Relative positions of floral parts (cf. p. 8). Type IV: (*a*) *Prunus*, (*b*) *Geum*, (*c*) *Daphne*; type V: (*d*) *Saxifraga stolonifera*, (*e*) *Campanula*. A – androecium, C – corolla, G – gynoecium, K – calyx, P – perianth (undifferentiated), pz – perigynous zone (shaded).

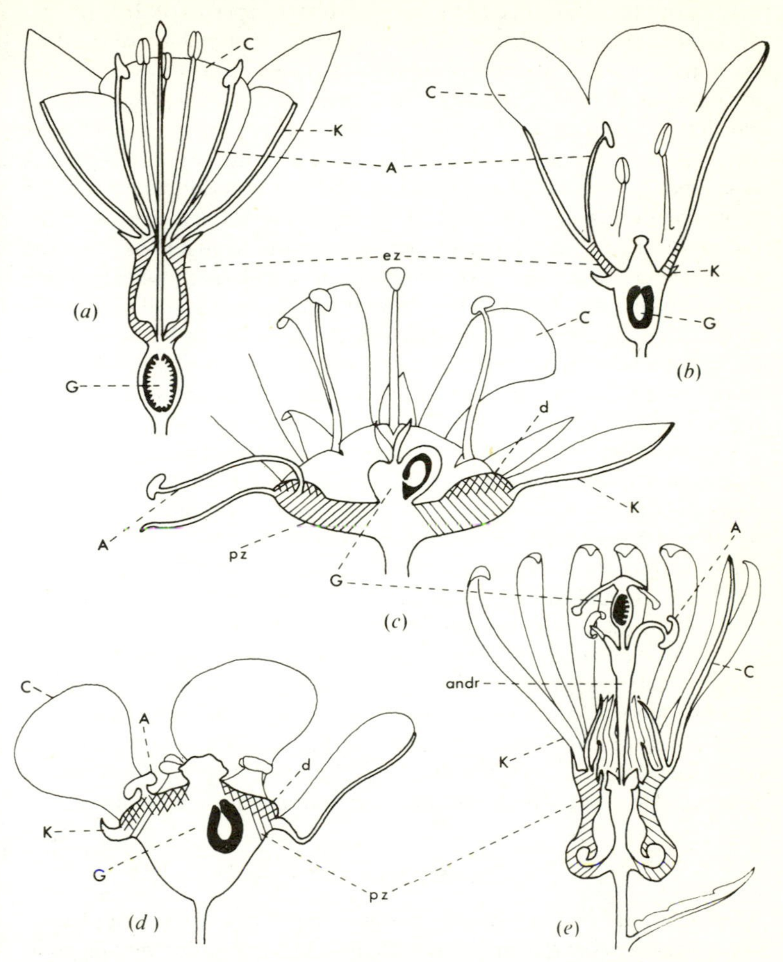


Figure 4. Relative positions of floral parts (cf. pp. 8–12). Type VI: (*a*) *Fuchsia*, (*b*) *Viburnum*. More complicated types: (*c*) *Acer* (cf. p. 12), (*d*) *Euonymus* (cf. p. 12), (*e*) *Passiflora* (cf. p. 12). A – androecium, andr – androgynophore, C – corolla, d – disc (cross-hatched), ez – epigynous zone (hatched), G – gynoecium, K – calyx, pz – perigynous zone (hatched).

the edge of it, and may therefore be said to be perigynous. The stamens may be borne on the edge of the disc, as in many species of *Acer* (cf. fig. 4(*c*)) when they are considered as perigynous, or on the top of the disc, as in many species of *Euonymus* (fig. 4(*d*)) when they are treated as hypogynous. A further complication appears in *Passiflora* (fig. 4(*e*)), in which the ovary and stamens are elevated on an androgynophore. In the key, *Passiflora* is placed in group 4 because only the perianth is perigynous, the stamens being hypogynous. Another situation which may be confusing occurs in *Silene* (fig. 2(*b*)) and other related genera of the *Caryophyllaceae*, in which all the whorls are hypogynous, but the corolla, androecium and ovary are borne on a stalk (anthophore) which is usually enclosed in the calyx. It must also be mentioned that transitions occur between the types described, particularly between types I–IV; in some cases it is very difficult to decide how a particular flower is to be treated, and careful observation is absolutely essential. To achieve this a longitudinal section of the flower must be cut.

In conclusion it must be stressed that we have used these terms with reference to flowers at anthesis; in later stages the relationship of the parts often changes. However, it is usually possible to tell that a fruit has developed from a superior or inferior ovary by the positions of the scars – if the scars of the floral parts surround and are contiguous with the stalk the ovary was superior (e.g. *Papaver*); if, however, they are near the top of the fruit, the ovary was inferior (e.g. *Malus*).

We have followed tradition in not applying the hypogyny, perigyny, epigyny terminology to the monocotyledons. In the descriptions of the Dicotyledonous families the terms (abbreviated to 'hypog', 'perig' and 'epig') are used; the ovary position is not normally given, but may be easily found by reference to table 1 (p. 6) and fig. 1 (p. 7). The ovary is never inferior unless both perianth and stamens are epigynous.

## PLACENTATION

Although there is some overlap between the different types of placentation, we may conveniently deal with this character under the following headings, adopting the terms as they are used in the key. Placentation should be observed in both transverse and longitudinal sections of the ovary.

1. MARGINAL. This is used only of free carpels, and describes the condition in which the carpel bears several ovules on its adaxial margin (e.g. *Caltha*, *Pisum*, etc.; cf. fig. 5(*a*), (*b*)).

2. AXILE. Here the ovary is always syncarpous and septate. The

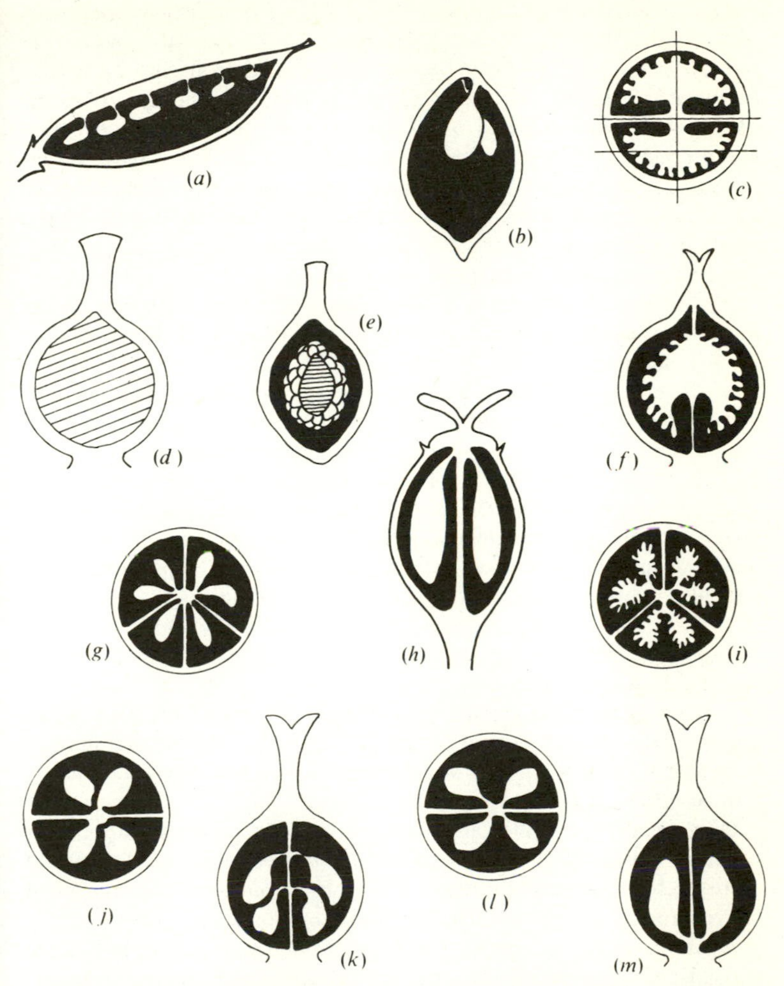


Figure 5. Placentation (cf. pp. 12–14). (*a*), (*b*) *Marginal*; (*c*)–(*m*) *Axile*. (*c*)–(*f*) ovules on swollen placentas ((*c*) transverse, (*d*)–(*f*) longitudinal sections; planes of the longitudinal sections indicated in (*c*)). (*g*) ovules borne on the axis. (*h*) ovules pendulous. (*i*) ovules on intrusive placentas. (*j*), (*k*) ovules superposed. (*l*), (*m*) ovules side by side, ascending.

ovules are borne on the central axis (e.g. *Narcissus*; cf. fig. 5(*g*)), on swollen placentae (e.g. *Solanum*; cf. fig. 5(*c*)–(*f*)) or intrusive placental outgrowths (*Begonia*; cf. fig. 5(*i*)). In some families the ovules are reduced to 1 or 2 in each loculus and ascend from the base (e.g. *Ipomoea*; cf. fig. 5(*l*), (*m*)), or are pendulous from the apex (*Umbelliferae*; cf. fig. 5(*h*)). Ovules in axile ovaries sometimes occur side by side (collateral), as in *Heliotropium* (cf. fig. 5(*l*), (*m*)), or superposed (*Acanthaceae*; cf. fig. 5(*j*), (*k*)). Occasionally, axile ovaries are further divided by secondary septa which grow inwards from the carpel walls as the ovary matures (e.g. *Linum*, *Salvia*), so that the ovary comes to have twice as many loculi as carpels.

3. PARIETAL. This term is used when the ovules are borne on the *walls* of the ovary, or on outgrowths from them. Several situations may be distinguished.

In the majority of cases parietal placentation occurs in unilocular, syncarpous ovaries, the ovules being confined to restricted placental regions on the walls (usually interpreted as carpel margins), as in *Viola* (cf. fig. 6(*a*)), *Gentiana* (cf. fig. 6(*d*)–(*f*)) or *Ribes*, or on intrusive, placenta-bearing outgrowths from them (e.g. *Cistus*, *Heuchera*; cf. fig. 6(*b*)). Intrusive parietal placentas may nearly meet in the middle of the ovary, so that the distinction between axile and parietal placentation is not always clear-cut (e.g. *Escallonia*, *Cucumis*; cf. fig. 6(*g*)–(*i*)). It may, in fact, appear to change during the development of the ovary (e.g. *Mesembryanthemum*); in other cases, an ovary may be axile below, parietal above (e.g. *Deutzia*).

In some syncarpous groups the ovules are borne on the walls of an apparently 2- or more-locular ovary (cf. fig. 6(*c*)). The septum may be thick (e.g. *Glaucium*) or thin and membranous (most *Cruciferae*), and is termed a false septum or replum.

Occasionally the ovules are scattered over the carpel surfaces. This situation is usually distinguished as diffuse-parietal placentation, and occurs in both apocarpous gynoecia (e.g. *Butomaceae*; cf. fig. 6(*j*)) and in syncarpous ovaries (*Hydrocharitaceae*; cf. fig. 6(*k*)).

As shown in the accompanying diagrams, the side view of both axile and parietal placentation can vary greatly according to the vertical plane in which the ovary has been cut. It can best be understood in relation to a transverse section.

4. FREE-CENTRAL. The ovules (usually many) are borne on a central globose or columnar structure that rises from the base of a syncarpous, unilocular ovary (e.g. *Pinguicula*; cf. fig. 7(*a*)–(*c*)). In most cases a thread of non-ovuliferous tissue attaches the placental column to

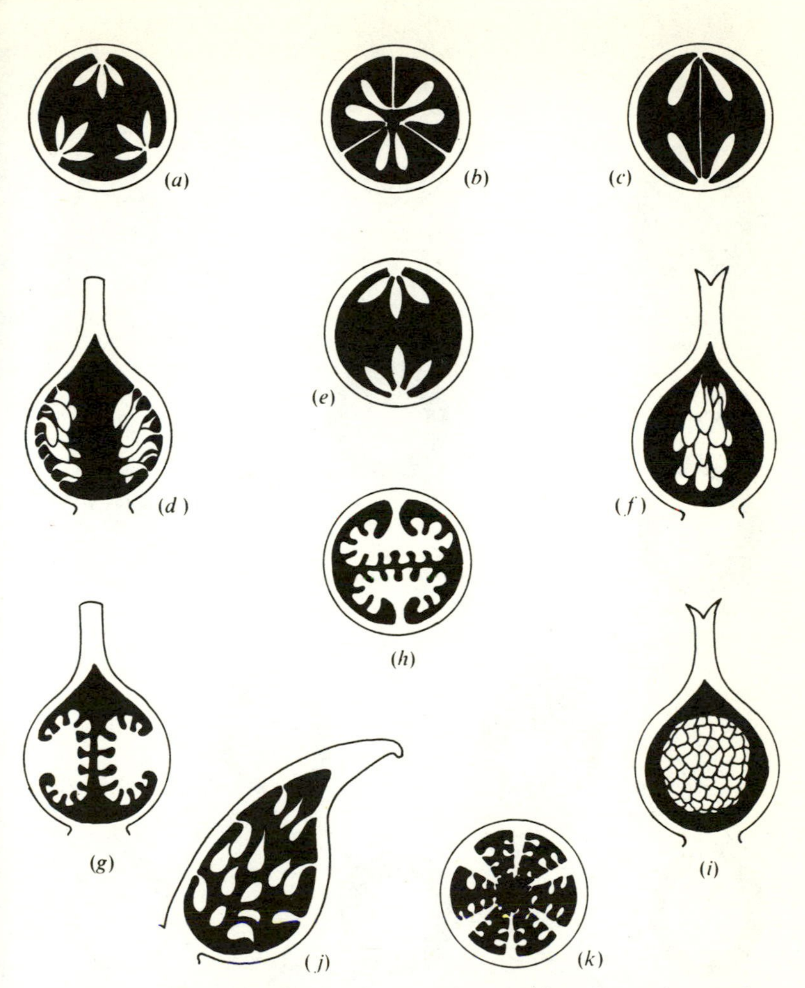


Figure 6. Placentation (cf. p. 14). *Parietal types*. (*a*) ovules on the carpel walls. (*b*) ovules on intrusive placentas. (*c*) ovules on the carpel walls, septum present. (*d*)–(*f*) ovules on the carpel walls: (*d*) longitudinal section through placentas, (*e*) transverse section, (*f*) longitudinal section at right angles to placentas. (*g*)–(*i*) ovules on intrusive placentas which almost meet in the centre of the ovary: (*g*) longitudinal section through the placentas, (*h*) transverse section, (*i*) longitudinal section between the placentas. (*j*), (*k*) diffuse parietal.

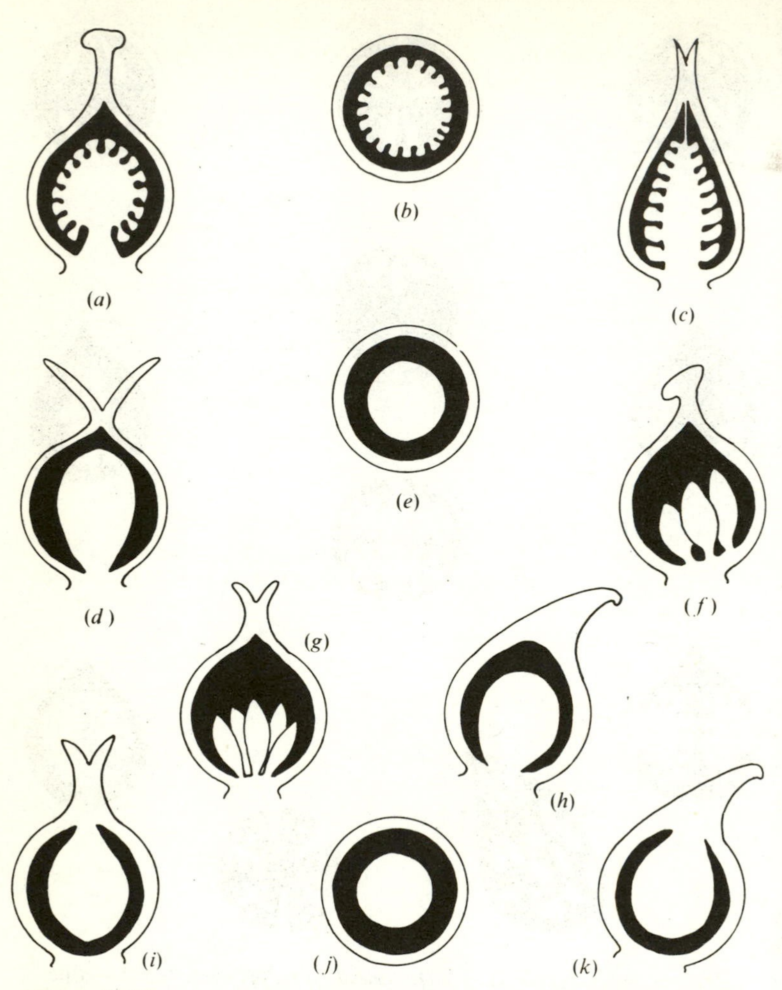


Figure 7. Placentation types (cf. pp. 14–17). (*a*)–(*c*) *Free-central*; (*d*)–(*h*) *basal*; (*i*)–(*k*) *apical*. (*a*), (*b*) ovules free-central. (*c*) ovules free-central showing attachment of placenta to the top of the ovary. (*d*), (*e*) one basal ovule. (*f*) ovules on an oblique placental cushion. (*g*) several basal ovules (ovary syncarpous). (*h*) one basal ovule (ovary apocarpous). (*i*), (*j*) ovule apical (ovary syncarpous). (*k*) ovule apical (ovary apocarpous).

the top of the ovary; sometimes this thread is rather stout (e.g. *Lysimachia*; cf. fig. 7(*c*)). Occasionally the ovary remains septate below (e.g. *Silene*), although in most ovaries with free-central placentation the septa break down as the ovary matures.

5. BASAL. Here the ovules arise from the base of a unilocular ovary (e.g. *Polygonum*, *Tamarix*, *Armeria*; cf. fig. 7(*d*), (*e*), (*g*)) or are borne on a basal placental cushion (oblique in *Berberis*; cf. fig. 7(*f*)). The term may be applied to both syncarpous ovaries (e.g. *Polygonum*) and free carpels (e.g. *Ranunculus*; cf. fig. 7(*h*)). There is no clear dividing line between basal and free-central placentation.

6. APICAL. In this case the ovule is attached to the apex of the single loculus, as in *Scabiosa* (syncarpous; cf. fig. 7(*i*), (*j*)) or *Anemone* (apocarpous; cf. fig. 7(*k*)). Although it would be possible to describe the solitary ovules in the loculi of many septate ovaries as apical or basal, we have not used the terms for those situations, to avoid confusion. 'Pendulous' or 'ascending' have been used instead and are treated as forms of axile placentation.

## AESTIVATION

The relationships of the lobes of the perianth in bud, known as aestivation, is another useful character that sometimes causes difficulty. In the key, three basic types are used; these are illustrated in fig. 8. The character is best observed in a transverse section of a bud just before opening, but may often be deduced from the positions of the bases of the parts (petals, sepals or perianth segments) of a mature flower. One rarely occurring type, which is not illustrated, is seen in the petals of *Papaver* and *Lythrum*, which are irregularly crumpled.

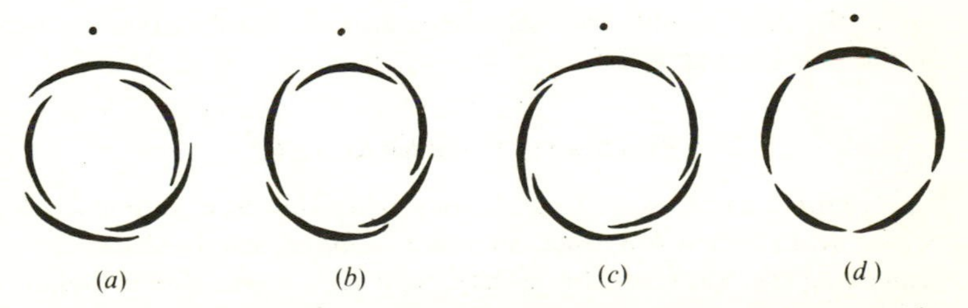


Figure 8. Aestivation types. (*a*), (*b*) imbricate (the details of the manner in which the organs overlap each other is variable – two of the commoner types are shown). (*c*) contorted. (*d*) valvate.

# EXAMINING THE PLANT

Using the keys provided in this book, the only tools usually essential for identifying an unknown flowering plant are a good hand-lens and a razor blade. If working indoors, a pair of dissecting needles and a dissecting stage are desirable; a low-power dissecting microscope may be resorted to in critical cases.

It is assumed that the student has some familiarity with plant morphology – that he will give particular attention to the arrangement of parts and to their positions of attachment, and that he will count the floral parts when these are not numerous. A flower is usually examined from the outside inwards, the parts being counted before the flower is sectioned. If the flower is zygomorphic (bilateral), it should be sectioned in the plane of symmetry.

In the following paragraphs, attention is mainly drawn to features that may cause difficulty when using the key, either because they are readily overlooked or difficult to interpret. The previous section should be consulted for placentation and the relative positions of the floral parts. It is assumed that the student has reasonably complete specimens, with vegetative parts, flowers of both sexes (if not bisexual) and fruits. In the keys we have tried to use rootstock and fruit characters only when other (generally more readily available) vegetative or floral characters are insufficient for identification.

## VEGETATIVE CHARACTERS

*Leaf position and blade.* In deciduous plants flowering without their leaves, leaf position (opposite, alternate, whorled) can be deduced by examining the leaf-scars. In the key, 'opposite' leaves should usually be interpreted as including the whorled condition. We have used the term 'divided' to apply to any deep division of a simple (i.e. not strictly compound) leaf. For the use of the term 'herb', see the glossary.

*Stipules.* Stipules may be present (often minute) or absent; normally at the base of the leaf blade or petiole, they may stand between two opposite leaves (interpetiolar stipules), in which position they are often united, or rarely between the leaf base and the stem (intrapetiolar stipules); the last two conditions occur in the *Rubiaceae*. It is important to remember that stipules are often deciduous; they should be looked for on young shoots, or an older shoot should be examined for stipular scars.

Other distinctive features to look for include punctate pellucid glands (best seen when a leaf is held up to the light) which may or may not be aromatic when the leaf is crushed, milky or coloured sap, and distinctive indumentum (e.g. lepidote scales, medifixed or stellate hairs).

## FLORAL CHARACTERS

Except for the way in which the perianth is arranged in the bud (see previous section), floral characters, including placentation, should be observed at full anthesis, since this is the stage in which they are recorded in the key. Although living material will preferably be used, floral structure can also be examined if a pressed flower is boiled in water for a minute or two; it can then be dissected out in a drop of water under a low-power dissecting microscope.

Observe whether a flower is bisexual (hermaphrodite) or unisexual (dioecious or monoecious); in some cases styles and even non-functional ovaries (pistillodes) may be present in male flowers (*Ilex*), or non-functional stamens in the female (*Acer*). An actinomorphic flower has sometimes only two planes of symmetry (*Cruciferae*, most *Papaveraceae*, some *Fumariaceae*), but usually more. A zygomorphic flower has only one plane of symmetry. These two terms, denoting type of symmetry, may apply to the whole flower or to parts of it (particularly the corolla). Zygomorphy usually results from unequal growth in any floral whorl or fusion of some of its parts, from the stamens being less numerous than the corolla segments, or even from the deflection of the floral whorls or the growth of a nectariferous disc to one side. By convention the gynoecium is not taken into consideration when deciding whether a flower is actinomorphic or zygomorphic. A few flowers (e.g. *Canna*) have no plane of symmetry, and are termed asymmetric.

*Perianth.* It is important to observe whether the perianth is in one or more series, or absent; if seriate, whether the series resemble one another in form, colour and texture. If the latter is the case, or the perianth is lacking, the plant must be keyed out in Groups 1, 6, 7, 8 or 13.

When both calyx and corolla are present (i.e. the perianth is differentiated), the plant may fall into any group except 6, 7 or 8.

A warning is needed concerning the calyx. This may be caducous, being thrown off as the flower opens (e.g. *Papaver*), so that the mature flower appears to have a uniseriate perianth. In some genera, especially those with condensed inflorescences and inferior ovaries, the calyx teeth are minute, or the calyx is reduced to a rim, as in many *Umbelliferae*, *Rubiaceae* and *Compositae*. In such cases the flower is treated in the key as having a distinct calyx and corolla, not as having a uniseriate perianth. The latter may sometimes be conspicuous, a tubular, petaloid structure (as in many *Thymelaeaceae* and *Nyctaginaceae*), and must not be confused with the corolla of a genus with a reduced calyx. In some cases an epicalyx may be present outside the true calyx (e.g. *Geum*, *Hibiscus*); in others the perianth may bear a petaloid or filamentous outgrowth called a corona (*Narcissus*, *Passiflora*); coronas may also be of staminodial origin.

It is sometimes difficult to see if all the petals are united at the base. In gamopetaly the basal zone of union may be very short, and can most easily be observed by gently pulling off the whole corolla; by this means it is also possible to see whether the stamens are attached to the corolla or free from it. Several of the petals may be united in Groups 2, 3, 4 and 5, but it is only in Groups 10, 11 and 12 that they are all united at the base, usually in a ring or tube.

A nectar disc, often forming a ring, is present in many insect-pollinated flowers – see p. 8 for discussion of such discs. In others, nectar is secreted by separate glands (e.g. *Geranium*), or even by the perianth itself, often in sacs or spurs developed on the corolla (e.g. *Delphinium*). In *Malvaceae* it is secreted by nectaries borne on the calyx.

A valuable family character is sometimes provided by the aestivation of the calyx or corolla, which is discussed on p. 17.

*Androecium*. In a few families the stamens are equal in number and opposite to (i.e. on the same radii as) the petals (i.e. they are antipetalous). This is an almost constant feature of the *Berberidaceae*, *Lardizabalaceae*, *Sabiaceae*, *Rhamnaceae*, *Vitaceae*, *Myrsinaceae*, *Theophrastaceae*, *Primulaceae*, *Plumbaginaceae* and *Tecophilaeaceae*. In the condition known as obdiplostemony (e.g. *Geranium*) the stamens of the outer whorl are opposite the petals. In some groups the anthers, instead of opening by longitudinal slits, open by terminal pores. Such poricidal dehiscence is characteristic of the following families: *Ochnaceae*, *Elaeocarpaceae*, *Tremandraceae*, *Polygalaceae*, *Melastomataceae*, *Actinidiaceae* (part), *Ericaceae*, *Mayacaceae* and

*Tecophilaeaceae*, but occurs sporadically in other families. Occasionally anthers open by flap-like, lateral valves (e.g. *Lauraceae* and many *Berberidaceae*, *Monimiaceae* and *Hamamelidaceae*); this condition is known as valvular dehiscence.

Pollen is sometimes shed in tetrads (*Juncaceae*, *Droseraceae*, most *Ericaceae*); the form of the grain is often helpful in distinguishing a family from its allies.

*Gynoecium*. The different kinds of placentation, of vital importance with the group keys, have already been explained (pp. 12–17). Placentation is best observed by cutting one ovary transversely and another longitudinally with a razor blade or sharp scalpel; in critical cases the transverse section should be examined under a dissecting microscope. Free-central placentation is characteristic of the following families: *Caryophyllaceae*, *Portulacaceae*, *Myrsinaceae*, *Theophrastaceae*, *Primulaceae* and *Lentibulariaceae*, but occurs sporadically elsewhere.

With very small ovaries it is often helpful to choose a fertilised ovary that is beginning to swell, and to imbed it in pith before sectioning. Ovule attachment may sometimes be seen more easily by slitting up the ovary wall with a sharp needle than by trying to section it (e.g. *Compositae*, *Cyperaceae*). The number of ovules (though not their attachment) can often be seen by gently squeezing the cross-sectioned ovary so that the ovules pop up. It should be noted that the number of ovules is often greatly reduced after fertilisation, even to a single seed.

Difficulty is sometimes found in deducing the number of carpels involved in a syncarpous ovary, or indeed whether an ovary is syncarpous or represents a solitary free carpel. Apart from developmental and anatomical studies, which may raise as many questions as they solve, morphological information from various sources helps to solve the problem.

1. The number of styles or stigmas is usually a reliable indicator, and the one that is most widely applicable. Divided stylar arms may mislead one into thinking that there are more carpels than there really are (e.g. *Euphorbia*). No help can be derived from stigma number if the stigma is capitate or punctate.

2. Marginal placentation always indicates a free carpel (e.g. *Pisum*). The number of placentae in a (syncarpous) ovary with axile or parietal placentation corresponds with the number of carpels, although the correlation may be disturbed by placental modifications (e.g. *Orobanchaceae*). In ovaries with free-central, basal or apical placentation, the position of the ovules is of no help.

3. A single, complete septum and 2 loculi indicates a bicarpellate

ovary (e.g. *Penstemon*). When more than 1, the number of septa or loculi can be misleading; the ingrowth of secondary septa from the wall in an ovary with axile placentation results in the ovary having twice as many septa and loculi as it has carpels (e.g. *Linum*, *Labiatae*). Sometimes only 1 of the loculi is fertile, the others being much reduced (e.g. *Viburnum* and many *Valerianaceae*); in such cases the number of stigmas often indicates the basic carpel number.

4. In capsular fruits, the fruit often splits into valves that correspond to carpel units or, when loculicidally dehiscent, into adjacent half-carpels. However, some capsules may have twice as many teeth as carpels, due to splitting of the carpel tips (e.g. *Silene*). In schizocarpic fruits, the number of mericarps corresponds to carpel number (e.g. *Umbelliferae*, *Geranium*), unless there has been false septation (e.g. *Labiatae*). With indehiscent and non-schizocarpic fruits such as berries, drupes and nuts there is, of course, no information to be derived from such sources.

In the keys and descriptions, when an ovary is described as having a certain number of loculi (cells) without any mention of the carpel number, this is generally because the latter is uncertain, or the number of loculi easier to see. Indeed, in much modified, unilocular ovaries (e.g. *Pandanus*, *Mirabilis*, *Berberis*) estimation of carpel number largely depends on the interpretation of vascular anatomy.

# KEYS

## USING THE KEYS

The keys given here are of the bracketed type, and are dichotomous throughout, i.e. at every stage a choice must be made between two (and only two) contrasting alternatives, which together make up a 'couplet'. As the main key allows for the identification of 256 families, it has been arranged in groups, and thus consists of 15 parts: a key to the groups and 14 group keys. To facilitate reference to particular leads (i.e. individual alternative statements), each couplet is numbered, and each lead is given a distinguishing letter (a or b). Three families are given in a very broad sense in the main key; keys to their segregate families (of which 16 are recognised) are given on pp. 59–61. These broad families may be recognised in the main key by the letters 's.l.' after their names (s.l. standing for *sensu lato* – in the broad sense).

To find the family to which a specimen belongs, one starts with the key to the groups, and compares the specimen with the two alternatives of the couplet numbered '1' in this key. If the plant agrees with 1a, one proceeds to the couplet numbered 2; if it agrees with 1b, to the couplet numbered 3. The process described above is then repeated for the subsequent couplets until a lead indicating a particular group is reached. It is very important that the whole of each couplet is read; otherwise wrong identifications may be made.

One proceeds in exactly the same way in the group keys until the name of a family is reached. The families are numbered in the key, and, to provide a check on the identification obtained, the families are briefly described, in numerical sequence on pp. 64–96. The specimen should be carefully compared with the description of the family; this should help to reveal errors in identification and observation.

In order that back-checking should be easy, the number of the lead from which any particular couplet has been derived is given in brackets after the couplet number of the 'a' lead. Thus '13a. (3)' means that one lead of the couplet numbered 3 takes one directly to the couplet numbered 13.

It will sometimes happen that the specimen does not agree with all the characters given in a particular lead. When this situation arises one must decide which of the two alternatives of the relevant couplet the specimen agrees with most fully. In general, we have put the most reliable diagnostic characters at the beginnings of the leads, so these characters should be observed with particular care. The only exception to this procedure occurs when the second of the two alternatives of a couplet reads 'Combination of characters not as above' (e.g. 9b in Group 2). In these cases the specimen must agree with all the characters given in the first ('a') alternative before it can be considered as a member of the family keying out at that lead; if it deviates in one or more characters from those detailed in the 'a' alternative, it must be treated as falling in the 'combination of characters not as above' category.

It is very important that the key to the groups be used with especial care, as completely wrong results will be obtained from trying to run down a specimen in the wrong group. When turning to the group keys, short descriptions are given at the beginning of each; these should always be used as a check.

In a few cases it has been impossible (in a work of this size) to detail all the differences between two very similar families when none of these differences is completely diagnostic. In these cases it may be necessary to consult some of the works mentioned in the section on further identification (pp. 97–100) before the specimen can be identified with certainty. We have tried to make the key as 'watertight' as possible, but, in spite of our efforts, it is certain that such cases will occur, particularly with exotic species unfamiliar to us. We have been fortunate in having the resources of an excellent Botanic Garden and herbarium at our disposal, and have checked as many species as we could; we trust, therefore, that these difficult cases will not be very numerous.

## KEY TO GROUPS

1a. Monocotyledons: cotyledon 1, terminal, or undifferentiated; leaves usually parallel-veined, but veins connected by cross-veinlets; leaves exstipulate, opposite only in some aquatics; flowers usually 3-merous; vascular bundles scattered in stem, lacking cambium; mature root system wholly adventitious 2

1b. Dicotyledons: cotyledons usually 2, lateral; leaves usually net-veined, but ultimate veinlets with free endings; leaves stipulate or not, alternate, opposite or whorled; flowers usually 2-, 4- or 5-merous or polymerous; vascular bundles of stem usually in a ring, with cambium; taproot usually persistent; 3

2a. (1) Ovary superior or naked; perianth well-developed, or reduced or absent; including all totally submerged aquatics **Group 13** (p. 53)

2b. Ovary fully or partly inferior; perianth well-developed, usually petaloid; if aquatic, never with flowers submerged at anthesis **Group 14** (p. 57)

3a. (1) At least some of the petals free at the base, or petals absent 4

3b. Petals present, all united at the base, sometimes shortly so 12

4a. (3) At least the male flowers borne in usually deciduous catkins; flowers always unisexual and apetalous; always woody **Group 9** (p. 44)

4b. Flowers not in catkins (or if so then plant herbaceous), polypetalous or apetalous, bisexual or unisexual; herbaceous or woody 5

5a. (4) Gynoecium of 2 or more free carpels **Group 1** (p. 26)

5b. Gynoecium of 1 carpel or syncarpous (if subapocarpous then carpels united near the base, or by a common style) 6

6a. (5) Perianth of 2 or more whorls, calyx and corolla both present (calyx rarely obsolescent; excluding aquatic plants with minute, quickly caducous petals, and branch parasites with opposite, leathery leaves) 7

6b. Perianth of 1 whorl, sometimes petaloid or 0; if perianth of 2 or more whorls then the segments of each whorl ± indistinguishable in flower 10

7a. (6) Stamens more than 2 × the number of petals (petals never more than 10) **Group 2** (p. 28)

7b. Stamens 2 × the number of petals or fewer 8

8a. (7) At least one whorl of the perianth hypogynous, or stamens hypogynous or inserted on the top of a hypogynous disc in which the ovary may be immersed 9

8b. Stamens and perianth perigynous, or ovary fully or partly inferior **Group 5** (p. 35)

9a. (8) Placentation axile, apical, basal or free-central **Group 3** (p. 31)

9b. Placentation parietal **Group 4** (p. 34)

10a. (6) Stamens not borne on the perianth or ovary naked 11
10b. Stamens apparently borne on the perianth, or ovary partly or fully inferior (female flowers sometimes without a perianth) **Group 8** (p. 41)
11a. (10) Flowers unisexual **Group 6** (p. 38)
11b. Flowers bisexual **Group 7** (p. 40)
12a. (3) Ovary superior 13
12b. Ovary partly or fully inferior **Group 12** (p. 51)
13a. (12) Corolla actinomorphic **Group 10** (p. 45)
13b. Corolla zygomorphic **Group 11** (p. 49)

## DICOTYLEDONS

### Group 1

*All of the petals free at the base, or petals absent; gynoecium of 2 or more free carpels.*

1a. Trees with exfoliating bark, palmately lobed leaves and unisexual flowers in pendulous, globose heads **29. Platanaceae**
1b. Combination of characters not as above 2
2a. (1) Perianth and stamens hypogynous 3
2b. Perianth and stamens perigynous 23
3a. (2) Aquatic plants with peltate leaves and 3 sepals 4
3b. Terrestrial plants, or if aquatic then usually without peltate leaves and with more than 3 sepals 5
4a. (3) Carpels sunk in an obconical receptacle; perianth polymerous **17. Nelumbonaceae**
4b. Carpels not sunk in the receptacle; perianth 2-seriate **16. Nymphaeaceae**
5a. (3) Herbs and succulent shrubs (rarely non-succulent shrubs with yellow wood), if climbers then with bisexual flowers and opposite leaves 6
5b. Trees, shrubs (never succulent or with yellow wood), or if climbers then with unisexual flowers and alternate leaves 11
6a. (5) Perianth absent **11. Saururaceae**
6b. Perianth present 7
7a. (6) Leaves usually succulent; stamens in 1 or 2 whorls **105. Crassulaceae**
7b. Leaves not succulent; stamens spirally arranged, numerous and indefinite 8

8a. (7) Petals laciniate; fruits borne on a common gynophore **85. Resedaceae**

8b. Petals (when present) not laciniate, but sometimes modified for nectar secretion; fruits not borne on a common gynophore 9

9a. (8) Leaves opposite or whorled; flowers small, sessile, in axillary clusters; ovule 1, placentation basal **38. Phytolaccaceae**

9b. Combination of characters not as above. 10

10a. (9) Sepals heteromorphic, green; stamens centrifugal, inserted on a nectariferous fleshy disc **51. Paeoniaceae**

10b. Sepals homomorphic, green or petaloid, or if heteromorphic then perianth zygomorphic; fleshy disc absent **18. Ranunculaceae**

11a. (5) Leaves simple 12

11b. Leaves compound 21

12a. (11) Sepals and petals 5 13

12b. Perianth not as above 14

13a. (12) Leaves opposite; stamens 5–10 **152. Coriariaceae**

13b. Leaves alternate; stamens more numerous **50. Dilleniaceae**

14a. (12) Unisexual climbers 15

14b. Erect trees or shrubs, flowers usually bisexual 16

15a. (14) Carpels many; seeds not U-shaped **6. Schizandraceae**

15b. Carpels 3 or 6; seeds often U-shaped **21. Menispermaceae**

16a. (14) Stamens with connective truncate, over-topping anther sacs; fruit usually a fleshy syncarp; endosperm convoluted **3. Annonaceae**

16b. Stamens with connective not as above; fruit various, not as above; endosperm not convoluted 17

17a. (16) Carpels spirally arranged on an elongate torus; stipules large, caducous, leaving an annular scar **1. Magnoliaceae**

17b. Carpels in 1 whorl or abbreviated cluster; stipules 0, or minute or adnate to the petiole 18

18a. (17) Petals present 19

18b. Petals absent 20

19a. (18) Sepals imbricate, free, more than 6; ovules solitary **5. Illiciaceae**

19b. Sepals valvate or united, 2–6; ovules several **2. Winteraceae**

20a. (18) Leaves whorled; flowers bisexual; sepals minute or 0 **28. Eupteleaceae**

20b. Leaves opposite or alternate; flowers unisexual; sepals 4 **27. Cercidiphyllaceae**

21a. (11) Unisexual climbers, or erect shrubs with blue fruits; perianth 3-merous **20. Lardizabalaceae**

21b. Erect shrubs, fruits not blue; perianth not 3-merous 22
22a. (21) Flowers showy, bisexual; leaves not aromatic **51. Paeoniaceae**
22b. Flowers inconspicuous, unisexual; leaves aromatic **153. Rutaceae**
23a. (2) Flowers unisexual; leaves evergreen **7. Monimiaceae**
23b. Flowers bisexual; leaves usually deciduous 24
24a. (23) Stamens all fertile; perianth 4–9-merous; leaves usually alternate 25
24b. Inner stamens sterile; perianth polymerous; leaves opposite **8. Calycanthaceae**
25a. (24) Leaves exstipulate, entire; flowers solitary and terminal; seed with a divided aril **52. Crossosomataceae**
25b. Leaves usually stipulate and toothed (if not then flowers clustered); seed not arillate **107. Rosaceae**

## Group 2

*Calyx and corolla both present, at least some of the petals free at the base; stamens more than 2 × the number of petals; carpel 1 or ovary syncarpous, superior or inferior.*

1a. Perianth and stamens hypogynous 2
1b. Perianth and stamens perigynous or epigynous 27
2a. (1) Placentation axile or free-central 3
2b. Placentation marginal or parietal 18
3a. (2) Placentation free-central; sepals usually 2 **43. Portulacaceae**
3b. Placentation axile; sepals usually 5, rarely fewer 4
4a. (3) Leaves all basal, tubular, forming insectivorous pitchers; style peltately dilated **24. Sarraceniaceae**
4b. Leaves not as above; style not dilated 5
5a. (4) Leaves alternate 6
5b. Leaves opposite 16
6a. (5) Anthers opening by terminal pores 7
6b. Anthers opening lengthwise 9
7a. (6) Shrubs with simple, exstipulate leaves, often with stellate hairs; stamens inflexed in bud; fruit a berry **87. Actinidiaceae**
7b. Combination of characters not as above 8
8a. (7) Ovary often deeply lobed, borne on an enlarged torus or gynophore; petals not laciniate **53. Ochnaceae**
8b. Ovary not lobed, sessile; petals often laciniate **59. Elaeocarpaceae**

9a. (6) Inner whorl of perianth segments tubular or bifid, nectariferous; fruit a group of partly to fully coalescent follicles **18. Ranunculaceae**
9b. Combination of characters not as above 10
10a. (9) Leaves punctate with pellucid, aromatic glands **153. Rutaceae**
10b. Leaves not punctate thus 11
11a. (10) Stipules absent; leaves evergreen **54. Theaceae**
11b. Stipules present; leaves usually deciduous 12
12a. (11) Filaments free; anthers 2-celled 13
12b. Filaments united into a tube, at least around the ovary, often also around the styles; anthers often 1-celled 14
13a. (12) Disc absent; stamens more than 15; leaves simple **60. Tiliaceae**
13b. Disc present, conspicuous; stamens 15; leaves dissected, rarely simple, foetid **155. Zygophyllaceae**
14a. (12) Style 1, capitate or lobed, stigmas 1–several; pollen grains not spiny; stipules usually deciduous; carpels 2–5 15
14b. Styles divided, several; pollen grains spiny; stipules often persistent; carpels 5 or more **63. Malvaceae**
15a. (14) Stamens in 2 whorls, those of the outer whorl usually staminodial; trunks or stems relatively slender **61. Sterculiaceae**
15b. Stamens in several whorls, staminodes absent; trunks often swollen, spiny **62. Bombacaceae**
16a. (5) Sepals calyptrate; fruit separating into boat-shaped follicular units **101. Eucryphiaceae**
16b. Sepals not calyptrate; fruit not as above 17
17a. (16) Leaves simple, exstipulate, often with pellucid glands; stamens often united in bundles **58. Guttiferae**
17b. Leaves pinnate, stipulate, without pellucid glands; stamens not united in bundles **155. Zygophyllaceae**
18a. (2) Aquatic plants with cordate leaves; style and stigma forming a sessile disc **16. Nymphaeaceae**
18b. Combination of characters not as above 19
19a. (18) Carpel 1, with marginal placentation 20
19b. Carpels 2 or more, placentation parietal 21
20a. (19) Leaves bipinnate or phyllodic, stipulate (see p. 60) **108. Leguminosae** s.l.
20b. Leaves various, not as above, exstipulate **18. Ranunculaceae**
21a. (19) Leaves opposite 22
21b. Leaves alternate 24

22a. (21) Styles numerous; flowers 3-merous **22. Papaveraceae**
22b. Styles 1–5; flowers 4–5-merous 23
23a. (22) Style 1; stamens not united in bundles; leaves without pellucid glands **73. Cistaceae**
23b. Styles 3–5; stamens united in bundles; leaves with pellucid glands **58. Guttiferae**
24a. (21) Trees; leaves stipulate; anthers opening by pore-like slits **72. Bixaceae**
24b. Herbs or shrubs; leaves usually not stipulate; anthers opening lengthwise 25
25a. (24) Sepals 2(–3), quickly deciduous **22. Papaveraceae**
25b. Sepals 4–8, persistent in flower 26
26a. (25) Ovary borne on a gynophore, closed at apex; none of the petals laciniate **83. Capparaceae**
26b. Ovary sessile, open at apex; at least some of the petals laciniate **85. Resedaceae**
27a. (1) Flowers unisexual; leaf base oblique **79. Begoniaceae**
27b. Flowers bisexual; leaf base not oblique 28
28a. (27) Placentation free-central; ovary partly inferior **43. Portulacaceae**
28b. Placentation not free-central; ovary either superior or fully inferior 29
29a. (28) Aquatic plants with cordate leaves **16. Nymphaeaceae**
29b. Terrestrial plants; leaves various 30
30a. (29) Stamens united in antipetalous bundles; staminodes frequent; plants usually rough with bristly or stinging hairs **78. Loasaceae**
30b. Combination of characters not as above 31
31a. (30) Sepals 2 calyptrate; herbs **22. Papaveraceae**
31b. Sepals 4–5 not calyptrate; trees or shrubs 32
32a. (31) Carpels 8–12, superposed **118. Punicaceae**
32b. Carpels fewer ± side by side 33
33a. (32) Leaves stipulate 34
33b. Leaves exstipulate 35
34a. (33) Leaves opposite or whorled; plants woody **102. Cunoniaceae**
34b. Leaves alternate; plants woody or herbaceous **107. Rosaceae**
35a. (33) Leaves with pellucid aromatic glands (rarely obscure); style 1 **117. Myrtaceae**
35b. Leaves without pellucid aromatic glands; styles usually more than 1 (see p. 59) **106. Saxifragaceae** s.l.

## Group 3

*Calyx and corolla both present; at least some of the petals free at the base; at least 1 whorl of the perianth hypogynous, or stamens hypogynous or sometimes inserted on a hypogynous disc; stamens 2 × the number of petals or fewer; ovary superior, placentation axile, apical, basal or free-central.*

1a. Resinous trees or shrubs; leaves simple or pinnate, alternate; flowers with a hypogynous disc, mostly unisexual; stamens 4–5 or 8–10; fruit 1-seeded, drupe-like **149. Anacardiaceae**
1b. Combination of characters not as above 2
2a. (1) Placentation free-central (ovary sometimes septate below) or basal 3
2b. Placentation axile or apical 7
3a. (2) Stamens antipetalous 4
3b. Stamens not antipetalous 6
4a. (3) Anthers opening by valves; stigma 1 **19. Berberidaceae**
4b. Anthers opening by slits; stigmas more than 1 5
5a. (4) Sepals 5; ovule 1, basal on a long, curved funicle; leaves exstipulate **49. Plumbaginaceae**
5b. Sepals 2(–3); ovules usually numerous, when 1 not on a long, curved funicle; leaves stipulate **43. Portulacaceae**
6a. (3) Style gynobasic; leaves pinnatisect **161. Limnanthaceae**
6b. Style terminal: leaves entire **42. Caryophyllaceae**
7a. (2) Anthers opening by terminal pores; stamens not antipetalous 8
7b. Anthers opening by longitudinal slits or stamens antipetalous 13
8a. (7) Flowers strongly zygomorphic, papilionaceous 9
8b. Flowers actinomorphic, not papilionaceous 10
9a. (8) Stamens 8, filaments united for at least ½ their length; fruit without barbed bristles **166. Polygalaceae**
9b. Stamens 4 or 3, filaments not united; fruit covered with barbed bristles **167. Krameriaceae**
10a. (8) Ovary borne on an enlarged torus or gynophore; fruit separating into distinct drupes; leaves stipulate **53. Ochnaceae**
10b. Combination of characters not as above 11
11a. (10) Carpels 2; ovules 1–2 per loculus **165. Tremandraceae**
11b. Carpels 3 or 5; ovules numerous in each loculus 12

12a. (11) Ovary 3-locular; pollen not in tetrads; style divided above into 3 branches; disc absent **89. Clethraceae**

12b. Ovary (3–)5-locular; pollen mostly in tetrads; style undivided; disc usually present **90. Ericaceae**

13a. (7) Herbs 14

13b. Trees, shrubs or climbers 23

14a. (13) Petals and stamens numerous; plants with succulent leaves **41. Aizoaceae**

14b. Petals 4–5; stamens 4–10; leaves not usually succulent 15

15a. (14) Small glabrous annual herb growing in water or wet mud; seeds pitted **57. Elatinaceae**

15b. Combination of characters not as above 16

16a. (15) Leaves aromatic, glandular-punctate; ovary often borne on a gynophore **153. Rutaceae**

16b. Leaves neither aromatic nor glandular-punctate; gynophore usually absent 17

17a. (16) Fruit ± 10(–6)-locular due to 5(–3) false septa; leaves simple, entire **157. Linaceae**

17b. Fruit 2–5(–8)-locular; leaves simple, lobed or compound 18

18a. (17) Carpels 2 or 4 (see p. 59) **106. Saxifragaceae** s.l.

18b. Carpels 5 19

19a. (18) Disc present; leaves pinnate or 2–3-foliolate **155. Zygophyllaceae**

19b. Disc absent or inconspicuous; leaves simple or palmately lobed or pinnatisect, rarely trifoliolate 20

20a. (19) Leaves stipulate, usually palmately or pinnately lobed or compound; fruit often long-beaked **159. Geraniaceae**

20b. Leaves exstipulate; fruit not long-beaked 21

21a. (20) Perianth actinomorphic; leaves compound; styles 5, free **160. Oxalidaceae**

21b. Perianth strongly zygomorphic, spurred; leaves simple or divided; style 1 or obsolete 22

22a. (21) Leaves peltate, sometimes palmately divided; stamens 8; carpels 3; fruit of 3 mericarps **162. Tropaeolaceae**

22b. Leaves with basal petiole, simple; stamens 5; carpels 5; fruit a fleshy, explosive capsule **163. Balsaminaceae**

23a. (13) A well-developed hypogynous glandular disc present below and/or around the ovary 24

23b. Hypogynous disc absent 35

24a. (23) Stamens antipetalous 25

24b. Stamens not antipetalous 26

25a. (24) Inflorescence leaf-opposed; usually climbers with leaf-opposed tendrils **140. Vitaceae**
25b. Inflorescence not leaf-opposed; trees or shrubs without tendrils **144. Sabiaceae**
26a. (24) Leaves with pellucid aromatic glands **153. Rutaceae**
26b. Leaves without pellucid aromatic glands 27
27a. (26) Resinous trees; style 1; fruit a 1–5-seeded drupe **148. Burseraceae**
27b. Combination of characters not as above 28
28a. (27) Corolla zygomorphic 29
28b. Corolla actinomorphic 31
29a. Leaves opposite, palmate; sepals united at base **146. Hippocastanaceae**
29b. Leaves alternate, usually pinnate; sepals free 30
30a. (29) Stipules large, intrapetiolar **142. Melianthaceae**
30b. Stipules absent or small and inconspicuous, not intrapetiolar **145. Sapindaceae**
31a. (28) Flowers functionally unisexual or polygamous 32
31b. Flowers all bisexual 33
32a. (31) Leaves usually alternate; ovary 5–2-carpellate, not flattened **150. Simaroubaceae**
32b. Leaves opposite; ovary usually 2(–3)-carpellate, flattened in a plane at right angles to the septum **147. Aceraceae**
33a. (31) Leaves entire or toothed; stamens 4–5, emerging from the fleshy disc **133. Celastraceae**
33b. Combination of characters not as above 34
34a. (33) Leaves exstipulate, not fleshy; filaments usually connate into a tube **154. Meliaceae**
34b. Leaves stipulate, fleshy; filaments free **155. Zygophyllaceae**
35a. (23) Petals long-clawed, often fringed or toothed; some or all of the sepals with abaxial nectaries; hairs often medifixed **164. Malpighiaceae**
35b. Combination of characters not as above 36
36a. (35) Filaments united below 37
36b. Filaments free 38
37a. (36) Filaments clearly united; stipules persistent, intrapetiolar; petals appendaged **158. Erythroxylaceae**
37b. Filaments clearly to obscurely united; stipules caducous, not intrapetiolar; petals not appendaged **61. Sterculiaceae**
38a. (36) Stamens 2, anther cells back to back **174. Oleaceae**
38b. Stamens 3–10, anther cells not back to back 39

39a. (38) Stamens 8; ovary 4-locular; flowers subtended by a pair of basally connate bracteoles **55. Stachyuraceae**
39b. Combination of characters not as above 40
40a. (39) Stamens 10 or 5+5 staminodes; inflorescence racemose **88. Cyrillaceae**
40b. Stamens 5–3; staminodes absent unless flowers functionally unisexual; inflorescence cymose or flowers axillary 41
41a. (40) Stamens 3–4; petals 3–4, yellow; small shrub **151. Cneoraceae**
41b. Stamens and petals 5 or if 4 then petals white or greenish; trees, shrubs or climbers 42
42a. (41) Ovules 1–2 on each placenta; style very short or absent **135. Aquifoliaceae**
42b. Ovules numerous on each placenta; style well-developed **103. Pittosporaceae**

## Group 4

*Calyx and corolla both present; at least some of the petals free at the base; at least 1 whorl of the perianth hypogynous, or stamens hypogynous or sometimes inserted on a hypogynous disc; stamens 2 × the number of petals or fewer; ovary superior, placentation parietal or marginal.*

1a. Flowers zygomorphic 2
1b. Flowers actinomorphic 6
2a. (1) Ovary of 1 carpel with marginal placentation; fruit a legume, sometimes indehiscent or lomentoid (see p. 60) **108. Leguminosae** s.l.
2b. Ovary of 2 or more carpels, or if of 1 carpel then with almost basal placentation; fruit various, never a legume 3
3a. (2) Carpels open at apex; some or all of the petals laciniate **85. Resedaceae**
3b. Carpels closed at apex; no petals laciniate 4
4a. (3) Stamens and petals 5; carpels 3 **69. Violaceae**
4b. Stamens and petals 4 or 6; carpels 2 5
5a. (4) Ovary borne on a gynophore; stamens usually exserted **83. Capparaceae**
5b. Ovary usually sessile; stamens not exserted **23. Fumariaceae**
6a. (1) Parasites or saprophytes without chlorophyll **90. Ericaceae**
6b. Free-living plants with chlorophyll 7
7a. (6) Petals and stamens numerous 8

7b. Petals and stamens definite, usually less than 7 — 9
8a. (7) Climbers; leaves not succulent — **68. Flacourtiaceae**
8b. Not climbing; leaves succulent — **41. Aizoaceae**
9a. (7) Trees or shrubs with palmately veined leaves; corona present between petals and stamens, bearing staminode-like appendages — **143. Greyiaceae**
9b. Combination of characters not as above — 10
10a. (9) Stamens alternating with multifid staminodes (see p. 59) — **106. Saxifragaceae** s.l.
10b. Stamens not alternating with multifid staminodes — 11
11a. (10) Leaves insectivorous by means of glandular hairs — **104. Droseraceae**
11b. Leaves not insectivorous — 12
12a. (11) Climbers with tendrils; ovary and stamens borne on an androgynophore; corona present — **71. Passifloraceae**
12b. Combination of characters not as above — 13
13a. (12) Petals 4, the inner pair usually deeply trifid; sepals 2 — **22. Papaveraceae**
13b. Petals not as above; sepals 4–5 — 14
14a. (13) Stamens 2 + 4; carpels apparently 2; ovary divided by a false septum — **84. Cruciferae**
14b. Stamens 4–10; carpels 2–5; false septum absent — 15
15a. (14) Petals with ligular scale at the base of the limb; leaves opposite — **75. Frankeniaceae**
15b. Petals without a ligular scale; leaves alternate — 16
16a. (15) Leaves exstipulate, usually scale-like; corolla and stamens arising from a fleshy disc; seeds hairy — **74. Tamaricaceae**
16b. Leaves stipulate, not scale-like; corolla and stamens not arising from a fleshy disc; seeds not hairy — **69. Violaceae**

## Group 5

*Calyx and corolla both present, the former sometimes reduced to a rim; at least some of the petals free at the base; both stamens and perianth perigynous or epigynous; stamens 2 × the number of petals or fewer; ovary superior to inferior. (Excluding aquatic plants with minute caducous petals and branch parasites with opposite, leathery leaves.)*

1a. Petals and stamens numerous; plant succulent — 2
1b. Petals and stamens definite, usually less than 10; plants usually not succulent — 3

2a. (1) Stem succulents (rarely shrubs); often very spiny and with much reduced or obsolete leaves; placentation parietal **40. Cactaceae**

2b. Leaf succulents, without spines; placentation axile or parietal (when septate) **41. Aizoaceae**

3a. (1) Anthers poricidal; stamens geniculate; leaves with 3 parallel main veins **120. Melastomataceae**

3b. Anthers opening by slits or valves; stamens not geniculate; leaves with reticulate venation 4

4a. (3) Carpel 1 with marginal placentation; fruit a legume (see p. 60) **108. Leguminosae** s.l.

4b. Carpels more than 1, united, placentation not marginal; fruit never a legume 5

5a. (4) Placentation parietal, placentae sometimes intrusive 6

5b. Placentation axile, basal, apical or free-central 10

6a. (5) Climbers with tendrils; flowers unisexual **81. Cucurbitaceae**

6b. Herbs or shrubs without tendrils; flowers usually bisexual 7

7a. (6) Trees with 2–3-pinnate leaves; flowers zygomorphic; stamens 5, of different lengths; staminodes present **86. Moringaceae**

7b. Combination of characters not as above 8

8a. (7) Petals contorted; ovary of 3 carpels; stigmas brush-like **70. Turneraceae**

8b. Petals imbricate; ovary not 3-carpellate; stigmas various, not brush-like 9

9a. (8) Aquatic herb; stamens 2 + 4 **84. Cruciferae**

9b. Terrestrial herbs or shrubs; stamens 4–5 or 8–10 (see p. 59) **106. Saxifragaceae** s.l.

10a. (5) Placentation free-central 11

10b. Placentation axile, basal or apical 12

11a. (10) Sepals usually 2; capsule dehiscing by a lid **43. Portulacaceae**

11b. Sepals 4–5; capsule opening by teeth **42. Caryophyllaceae**

12a. (10) Stamens antipetalous; trees or shrubs with simple leaves **139. Rhamnaceae**

12b. Stamens antisepalous or 2 × as many as petals; herbaceous or woody, leaves simple to compound 13

13a. (12) Flowers borne in umbels, sometimes condensed into heads or superposed whorls; leaves usually compound; ovary inferior 14

13b. Flowers usually not borne in umbels; leaves usually simple; ovary superior or inferior 15

14a. (13) Fruit a schizocarp splitting into 2 mericarps; flowers usually bisexual; petals imbricate; mostly herbs without stellate hairs **169. Umbelliferae**

14b. Fruit a berry; flowers often unisexual; petals valvate; mostly woody, often with stellate hairs **168. Araliaceae**

15a. (13) Style 1 16

15b. Styles more than 1, often 2, divergent 26

16a. (15) Floating aquatic herb; petioles inflated **116. Trapaceae**

16b. Terrestrial herbs, shrubs or trees; petioles not inflated 17

17a. (16) Ovary 1-locular with 2–5 ovules; fruit leathery or drupe-like, 1-seeded **121. Combretaceae**

17b. Ovary (1–)2–5-locular, ovules various; fruit not as above 18

18a. (17) Ovules solitary in each loculus 19

18b. Ovules 2–numerous in each loculus 22

19a. (18) Petals valvate; flowers usually bisexual 20

19b. Petals imbricate; flowers often unisexual 21

20a. (19) Stamens with swollen, villous filaments; petals recurved **124. Alangiaceae**

20b. Stamens without swollen, villous filaments; petals not recurved **125. Cornaceae**

21a. (19) Flowers in heads, subtended by 2 conspicuous white bracts; ovary 10–6-locular **122. Davidiaceae**

21b. Flowers various, but not as above; ovary 1-locular **123. Nyssaceae**

22a. (18) Leaves glandular-punctate, aromatic; ovary borne on a gynophore **153. Rutaceae**

22b. Leaves neither aromatic nor glandular-punctate; ovary not borne on a gynophore 23

23a. (22) Ovary superior 24

23b. Ovary inferior 25

24a. (23) Calyx tube not prominently ribbed; seeds arillate; mostly trees, shrubs or climbers **133. Celastraceae**

24b. Calyx tube prominently ribbed; seeds not arillate; mostly herbaceous **114. Lythraceae**

25a. (23) Sap milky; petals 5; ovary 3-locular **199. Campanulaceae**

25b. Sap watery; petals 2 or 4; ovary (1–)4(–5)-locular **119. Onagraceae**

26a. (15) Fruit an inflated, membranous capsule; leaves mostly opposite, compound; stipules present **141. Staphyleaceae**

26b. Combination of characters not as above 26

27a. (26) Fruit a group of samaras; trees or shrubs with opposite, lobed or compound leaves **147. Aceraceae**
27b. Combination of characters not as above 28
28a. (27) Trees or shrubs; hairs often stellate; fruit a few-seeded woody capsule **30. Hamamelidaceae**
28b. Herbs or shrubs; hairs simple or absent; fruit various, not a woody capsule (see p. 59) **106. Saxifragaceae** s.l.

## Group 6

*Perianth of 1 whorl, sometimes petaloid or absent; if perianth of 2 whorls then the segments of each whorl ± indistinguishable; stamens not borne on the perianth; ovary superior or naked; flowers unisexual.*

1a. Aquatic herb; leaves with setaceous segments **15. Ceratophyllaceae**
1b. Terrestrial plants; leaves not as above 2
2a. (1) Trailing ericoid shrublet; fruit a berry **92. Empetraceae**
2b. Combination of characters not as above 3
3a. (2) Flowers in racemes or spikes; fruit a berry or drupe-like; leaves entire, alternate, exstipulate; carpels more than 5 **38. Phytolaccaceae**
3b. Combination of characters not as above 4
4a. (3) Ovary 3-locular; styles 3 5
4b. Ovary 1-, 2- or 4-locular; styles 1–2 7
5a. (4) Leaves with sheathing, membranous stipules; perianth segments 6; fruit a nut **48. Polygonaceae**
5b. Combination of characters not as above 6
6a. (5) Fruit schizocarpic; sap often milky; deciduous or evergreen herbs, shrubs or trees, or stem succulents; styles usually divided; seeds usually carunculate **137. Euphorbiaceae**
6b. Fruit a loculicidal capsule; sap not milky; evergreen shrubs; styles undivided; seeds black and shiny, not carunculate **136. Buxaceae**
7a. (4) Ovary 4-locular; leaves fleshy, opposite; female flowers without, male flowers with a perianth **47. Bataceae**
7b. Combination of characters not as above 8
8a. (7) Resinous trees or shrubs; leaves simple or pinnate; flowers with a disc; stamens 3–10; fruit 1-seeded, drupe-like **149. Anacardiaceae**
8b. Combination of characters not as above 9
9a. (8) Stamens 2, anther cells back to back **174. Oleaceae**

9b. Stamens more than 2, anther cells not back to back 10

10a. (9) Leaves formimg insectivorous pitchers **14. Nepenthaceae**

10b. Leaves neither insectivorous nor pitcher-like 11

11a. (10) Plants aromatic, dioecious; stamens 3–18, monadelphous; ovary of a single carpel with a single basal ovule **4. Myristicaceae**

11b. Combination of characters not as above 12

12a. (11) Placentation parietal; stamens numerous; fruit a berry or capsule **68. Flacourtiaceae**

12b. Combination of characters not as above 13

13a. (12) Trees, shrubs or climbers (rarely herbs, when flowers sunk in a fleshy receptacle); ovule apical 14

13b. Combination of characters not as above 17

14a. (13) Ovules 4, of which only 1 develops; flowers in axillary racemes **138. Daphniphyllaceae**

14b. Ovule 1; flowers not in axillary racemes 15

15a. (14) Sap watery; fruit a drupe **64. Ulmaceae**

15b. Sap milky; fruit a syncarp or samara 16

16a. (15) Perianth present; fruit frequently a syncarp of drupes or achenes + flat to flask-shaped receptacle **65. Moraceae**

16b. Perianth absent; fruit a samara **31. Eucommiaceae**

17a. (13) Stinging hairs present, or plant rough to the touch; stamens sensitive, inflexed in bud; often with cystoliths; seed with a straight embryo **67. Urticaceae**

17b. Stinging hairs absent; stamens neither sensitive nor inflexed; cystoliths absent; seed often with a curved embryo 18

18a. (17) Perianth scarious; stamens often connate below **46. Amaranthaceae**

18b. Perianth greenish or absent; stamens free 19

19a. (18) Leaves all opposite; fruit splitting into 2 mericarps **137. Euphorbiaceae**

19b. Leaves alternate, at least above; fruit not as above 20

20a. (19) Male flowers with 7–22 stamens; style at last lateral; leaf base sheathing **113. Theligonaceae**

20b. Combination of characters not as above 21

21a. (20) Ovary septate, 4-ovuled; leaves leathery **136. Buxaceae**

21b. Ovary unilocular, 1-ovuled; leaves not leathery 22

22a. (21) Leaves stipulate; ovule apical **66. Cannabaceae**

22b. Leaves exstipulate; ovule basal (including articulated stem succulents) **45. Chenopodiaceae**

## Group 7

*Perianth of 1 whorl, sometimes petaloid or absent; if perianth of 2 whorls then the segments of each whorl ± indistinguishable; stamens not borne on the perianth; ovary superior or naked; flowers bisexual. (Flowers rarely sunk in a spike-like receptacle.)*

1a. Flowers in racemes or spikes; fruit a berry or drupe-like; leaves entire, alternate, exstipulate **38. Phytolaccaceae**
1b. Combination of characters not as above 2
2a. (1) Trees or trailing ericoid shrublets, rarely an aromatic shrub 3
2b. Herbs, climbers or non-aromatic shrubs 9
3a. (2) Trailing ericoid shrublet; fruit a drupe **92. Empetraceae**
3b. Trees or non-ericoid shrubs; fruit a drupe, samara, nut or capsule 4
4a. (3) Stamens numerous; ovary 5- or more-locular 5
4b. Stamens 12 or fewer; ovary with fewer than 5 loculi 7
5a. (4) Leaves in whorls **26. Trochodendraceae**
5b. Leaves not in whorls 6
6a. (5) Flowers 4-merous, in catkin-like spikes **25. Tetracentraceae**
6b. Flowers not 4-merous, borne in cymes **60. Tiliaceae**
7a. (4) Leaves evergreen, with pellucid, aromatic glands; anthers opening by valves **9. Lauraceae**
7b. Leaves mostly deciduous, without pellucid, aromatic glands; anthers opening by slits 8
8a. (7) Stamens 2; leaf base not oblique **174. Oleaceae**
8b. Stamens 4–8; leaf base oblique **64. Ulmaceae**
9a. (2) Perianth absent; flowers borne (and often sunk) in a continuous fleshy spike; leaves well-developed, often fleshy **12. Piperaceae**
9b. Combination of characters not as above 10
10a. (9) Aquatics of running water, resembling algae, mosses or hepatics **109. Podostemaceae**
10b. Combination of characters not as above 11
11a. (10) Leaves stipulate (rarely apparently exstipulate), the stipules often united into a sheath (ochrea); fruit often a 3-sided nut **48. Polygonaceae**
11b. Leaves exstipulate; fruit not a 3-sided nut 12
12a. (11) Sepals caducous; tall herbs with palmately lobed leaves and orange sap **22. Papaveraceae**
12b. Combination of characters not as above 13
13a. (12) Ovary of 1 carpel, 1-seeded; perianth usually petaloid, bracts sometimes calyx-like **39. Nyctaginaceae**

13b. Ovary of 2 or more carpels, 1-many-seeded; perianth not petaloid 14
14a. (13) Carpels open at apex; placentation parietal **85. Resedaceae**
14b. Carpels closed at apex; placentation basal, free-central or axile 15
15a. (14) Ovule solitary, basal; leaves often alternate 16
15b. Ovules numerous on a free-central or axile placenta; leaves opposite or alternate 17
16a. (15) Perianth green, membranous or 0; stamens free **45. Chenopodiaceae**
16b. Perianth scarious; stamens often connate below **46. Amaranthaceae**
17a. (15) Placentation axile; leaves alternate (see p. 59) **106. Saxifragaceae** s.l.
17b. Placentation basal or free-central; leaves usually opposite 18
18a. (17) Sepals free; stamens opposite to, or more numerous than the sepals **42. Caryophyllaceae**
18b. Sepals united; stamens as many as, and alternating with, the sepals **100. Primulaceae**

## Group 8

*Perianth of 1 whorl, sometimes petaloid or absent; if perianth of 2 whorls then the segments of each whorl ± indistinguishable; stamens apparently borne on the perianth or ovary partly or fully inferior.*

1a. Aquatics or rhubarb-like marsh plants with cordate leaves 2
1b. Terrestrial plants, not as above 5
2a. (1) Leaves deeply divided, or very large and cordate; stamens 8, 4 or 2 3
2b. Leaves undivided; stamen 1 4
3a. (2) Ovary 2–4-locular; leaves exstipulate, deeply dissected **110. Haloragaceae**
3b. Ovary 1-locular; leaves stipulate, entire or lobed **111. Gunneraceae**
4a. (2) Leaves whorled; fruit a cypsela **112. Hippuridaceae**
4b. Leaves opposite; fruit 4-lobed **187. Callitrichaceae**
5a. (1) Trees or shrubs 6
5b. Herbs, climbers or parasites 18
6a. (5) Lepidote scales present; fruit enclosed in a berry-like calyx **127. Elaeagnaceae**
6b. Lepidote scales absent; fruit not as above 7

7a. (6) Stamen 1 median only, or 1 median and 2 lateral ½-stamens; leaves opposite **10. Chloranthaceae**
7b. Stamens not as above; leaves usually alternate 8
8a. (7) Stamens alternating with the sepals **139. Rhamnaceae**
8b. Stamens opposite to, or more numerous than the sepals 9
9a. (8) Ovary 2-locular, partly inferior; stellate hairs often present; fruit a woody capsule **30. Hamamelidaceae**
9b. Combination of characters not as above 10
10a. (9) Stamens 4, situated at the top of spoon-shaped, petaloid perianth segments which split apart as the flower opens **128. Proteaceae**
10b. Combination of characters not as above 11
11a. (10) Ovary inferior 12
11b. Ovary superior 15
12a. (11) Placentation parietal (see p. 59) **106. Saxifragaceae** s.l.
12b. Placentation axile or basal 13
13a. (12) Styles 3–6; fruit a nut surrounded by a scaly cupule **34. Fagaceae**
13b. Style 1; fruit not as above 14
14a. (13) Stamens 4–5; placentation basal **129. Santalaceae**
14b. Stamens 5–10; placentation axile **125. Cornaceae**
15a. (11) Leaves aromatic, dotted with pellucid glands; anthers opening by valves **9. Lauraceae**
15b. Leaves neither aromatic nor gland-dotted; anthers not opening by valves 16
16a. (15) Stamens 2, or 8–10 borne at different levels in the perianth tube; leaves simple, entire **115. Thymelaeaceae**
16b. Stamens not as above; leaves lobed or compound 17
17a. (16) Inflorescence borne on current year's shoots; fruit a schizocarp of 2(–3) samaras **147. Aceraceae**
17b. Inflorescence borne on old wood; fruit a legume (see p. 60) **108. Leguminosae** s.l.
18a. (5) Root or branch parasites 19
18b. Free-living plants 22
19a. (18) Branch parasites with forked branching, or flowers sessile on branches of host 20
19b. Root parasites, lacking chlorophyll 21
20a. (19) Flowers borne on green, forked branches **130. Loranthaceae**
20b. Flowers brown, minute, sessile **132. Rafflesiaceae**

21a. (19) Flowers minute in fleshy, crimson, phallus-like spikes; stamen 1 **131. Cynomoriaceae**

21b. Flowers conspicuous in showy, bracteate spikes; stamens 8 **132. Rafflesiaceae**

22a. (18) Perianth 0; flowers in spikes **11. Saururaceae**

22b. Perianth present; flowers usually not in spikes 23

23a. (22) Leaf base oblique; ovary inferior, 3-locular **79. Begoniaceae**

23b. Combination of characters not as above 24

24a. (23) Ovary superior 25

24b. Ovary partly or fully inferior 30

25a. (24) Carpel 1, ovule 1, apical; perianth tubular **115. Thymelaeaceae**

25b. Combination of characters not as above 26

26a. (25) Carpels (2–)3, ovule 1, basal; perianth persistent in fruit; leaves usually alternate, entire 27

26b. Combination of characters not as above 28

27a. (26) Leaves exstipulate; stamens 5 **44. Basellaceae**

27b. Leaves often with stipules (ochrea); stamens usually 6–9 **48. Polygonaceae**

28a. (26) Leaves opposite, usually entire 29

28b. Leaves alternate, usually lobed or compound **107. Rosaceae**

29a. (28) Stipules scarious, rarely absent; ovule 1; fruit a nut **42. Caryophyllaceae**

29b. Stipules greenish or absent; ovules numerous; fruit a capsule **114. Lythraceae**

30a. (24) Leaves pinnate; ovary open at apex **80. Datiscaceae**

30b. Leaves not pinnate; ovary closed 31

31a. (30) Ovary 6-locular; perianth 3-lobed or tubular and zygomorphic, bizarre **13. Aristolochiaceae**

31b. Combination of characters not as above 32

32a. (31) Ovules 1–5; seed 1 33

32b. Ovules and seeds numerous 34

33a. (32) Perianth segments thickening in fruit; leaves alternate **45. Chenopodiaceae**

33b. Perianth segments not as above; leaves opposite or alternate **129. Santalaceae**

34a. (32) Styles 2, placentation parietal (see p. 59) **106. Saxifragaceae** s.l.

34b. Style 1; placentation axile **119. Onagraceae**

## Group 9

*At least the male flowers borne in often deciduous catkins; flowers always apetalous and unisexual; always woody (seeds never with embryo ring-like).*

1a. Stems jointed; leaves reduced to whorls of scales **37. Casuarinaceae**
1b. Stems not jointed; leaves expanded 2
2a. (1) Leaves pinnate 3
2b. Leaves simple or lobed 4
3a. (2) Leaves exstipulate; fruit a nut **156. Juglandaceae**
3b. Leaves stipulate; fruit a legume (see p. 60) **108. Leguminosae** s.l.
4a. (2) Leaves opposite, evergreen, entire; fruit berry-like **126. Garryaceae**
4b. Leaves alternate, deciduous or evergreen; fruit not berry-like 5
5a. (4) Ovules many, parietal; seeds many, woolly; male catkins erect with the stamens exserted beyond the bracts, or pendulous with deciduous, laciniate bracts **82. Salicaceae**
5b. Ovules solitary or few; seeds few, glabrous; male catkins not as above 6
6a. (5) Leaves exstipulate, with golden, aromatic glands **33. Myricaceae**
6b. Leaves neither gland-dotted nor aromatic, stipulate or not 7
7a. (6) Styles 3, often divided; fruit schizocarpic; seeds carunculate **137. Euphorbiaceae**
7b. Styles 1–6, simple; fruit not schizocarpic; seeds not carunculate 8
8a. (7) Plants with milky sap **65. Moraceae**
8b. Plants without milky sap 9
9a. (8) Male catkins compound; flowers in groups of 2–3 adherent to their bract; styles 2 10
9b. Male catkins simple; flowers not grouped and adherent to a bract; styles 1 or 3–6 11
10a. (9) Nuts or samaras small, borne in 'cones'; perianth present in male flowers, absent in female flowers; ovary naked **35. Betulaceae**
10b. Nuts larger, subtended by a foliaceous bract or involucre; perianth present in female, absent in male flowers; ovary inferior **36. Corylaceae**

11a. (9) Ovary inferior; fruit a nut surrounded or enclosed by a scaly cupule; stipules deciduous; styles 3–6 **34. Fagaceae**

11b. Ovary superior; fruit a leather drupe, not enclosed in a cupule; stipules absent; style 1 **32. Leitneriaceae**

## Group 10

*Calyx and corolla both present, the petals all united, at least at the base; corolla actinomorphic; ovary superior.*

1a. Stamens 2, anther cells back to back; plants woody **174. Oleaceae**

1b. Combination of characters not as above 2

2a. (1) Carpels several, ± free; plants always succulent **105. Crassulaceae**

2b. Ovary syncarpous, or at least with the styles united, rarely 1-carpellate; plants seldom succulent 3

3a. (2) Parasites or saprophytes, without chlorophyll 4

3b. Free-living plants with chlorophyll 6

4a. (3) Twining stem parasites; corolla with scales below the attachment of the filaments **178. Cuscutaceae**

4b. Erect root parasites or saprophytes; corolla without scales 5

5a. (4) Stamens hypogynous; placentation parietal **90. Ericaceae**

5b. Stamens epipetalous; placentation axile **183. Lennoaceae**

6a. (3) Corolla scarious, 4-lobed; stamens 4, exserted; leaves often all basal and with parallel veins **188. Plantaginaceae**

6b. Combination of characters not as above 7

7a. (6) Bracts of the central, abortive flowers of the inflorescence modified into nectar-secreting pitchers; petals calyptrate, falling as the flower opens **56. Marcgraviaceae**

7b. Combination of characters not as above 8

8a. (7) Tropical trees or shrubs with milky sap and alternate leaves; stamens antipetalous or at least 2 × the number of corolla lobes 9

8b. Combination of characters not as above 10

9a. (8) Flowers bisexual; leaves entire; fruit a hard, few-seeded berry **94. Sapotaceae**

9b. Flowers mostly unisexual; leaves palmately lobed; fruit a large, many-seeded berry **77. Caricaceae**

10a. (8) Stamens antipetalous; placentation axile, free-central or basal 11
10b. Stamens antisepalous or more numerous or fewer than corolla lobes; placentation various, never free-central 16
11a. (10) Leaf-opposed tendrils usually present; petals calyptrate; placentation axile **140. Vitaceae**
11b. Tendrils absent; petals not calyptrate; placentation free-central or basal 12
12a. (11) Placentation basal; ovule 1, pendulous on long, curved funicle; fruit 1-seeded **49. Plumbaginaceae**
12b. Placentation free-central; ovules usually numerous though fruit often 1-seeded 13
13a. (12) Trees or shrubs; fruit a berry or drupe 14
13b. Herbs or woody only at base; fruit a capsule 15
14a. (13) Leaves glandular-punctate; anthers introrse; staminodes absent **99. Myrsinaceae**
14b. Leaves not glandular-punctate; anthers extrorse; staminodes 5 **98. Theophrastaceae**
15a. (13) Sepals 2, usually free **43. Portulacaceae**
15b. Sepals (4–)5(–9), connate **100. Primulaceae**
16a. (10) Flower compressed, with 2 planes of symmetry; stamens in 2 bundles **23. Fumariaceae**
16b. Combination of characters not as above 17
17a. (16) Leaves bipinnate or phyllodic; fruit a legume (see p. 60) **108. Leguminosae** s.l.
17b. Combination of characters not as above 18
18a. (17) Anthers opening by pores 19
18b. Anthers opening lengthwise, or pollen in masses (pollinia) 21
19a. (18) Stamens numerous **87. Actinidiaceae**
19b. Stamens 5, 8 or 10 20
20a. (19) Stamens hypogynous, often 2 × the number of corolla lobes **90. Ericaceae**
20b. Stamens epipetalous, usually 5, alternating with the corolla lobes **176. Solanaceae**
21a. (18) Stamens at least 2 × the number of corolla lobes 22
21b. Stamens as many as corolla lobes or fewer 26
22a. (21) Leaves with pellucid, aromatic glands; calyx cupular **153. Rutaceae**
22b. Leaves without pellucid, aromatic glands; calyx lobed 23
23a. (22) Leaves stipulate; stamens united, forming a long tube around the styles **63. Malvaceae**

23b. Leaves exstipulate; stamens free or shortly united 24
24a. (23) Spiny shrubs with fleshy leaves; placentation parietal **76. Fouquieriaceae**
24b. Woody but not spiny or fleshy; placentation axile 25
25a. (24) Flowers unisexual; fruit fleshy **95. Ebenaceae**
25b. Flowers bisexual; fruit a capsule **54. Theaceae**
26a. (21) Leaves alternate or all basal; carpels never 2 and almost free with single terminal style 27
26b. Leaves opposite or whorled, alternate only when carpels 2 and almost free with single terminal style 42
27a. (26) Woody, often evergreen; stigma sessile **135. Aquifoliaceae**
27b. Combination of characters not as above 28
28a. (27) Procumbent herbs with milky sap and hypogynous stamens **199. Campanulaceae**
28b. Combination of characters not as above 29
29a. (28) Ovary 5-locular 30
29b. Ovary 2-, 3- or 4-locular 31
30a. (29) Leaves fleshy; anthers 2-celled; fruit often deeply lobed, schizocarpic **175. Nolanaceae**
30b. Leaves leathery; anthers 1-celled; fruit a capsule or drupe **91. Epacridaceae**
31a. (29) Ovary 3-locular 32
31b. Ovary 2- or rarely 4-locular 33
32a. (31) Dwarf, evergreen shrublets; 5 staminodes usually present; corolla lobes imbricate **93. Diapensiaceae**
32b. Herbs, or climbers with tendrils; staminodes absent; corolla lobes contorted **180. Polemoniaceae**
33a. (31) Stamens syngenesious; flowers borne in capitula; stigma indusiate **201. Brunoniaceae**
33b. Combination of characters not as above 34
34a. (33) Flowers in scorpioid cymes or the calyx with appendaged sinuses; style gynobasic or terminal 35
34b. Flowers neither in scorpioid cymes (though often in dichasia) nor the calyx appendaged; style terminal 36
35a. (34) Style terminal; fruit a capsule, usually many-seeded **181. Hydrophyllaceae**
35b. Style usually gynobasic; fruit of 4 nutlets, rarely a 1–4-seeded drupe **182. Boraginaceae**
36a. (34) Placentation parietal 37
36b. Placentation axile (at least below) 38

37a. (36) Corolla lobes valvate; leaves glabrous, trifoliolate, cordate or peltate; aquatic or marsh plants **179. Menyanthaceae**
37b. Corolla lobes imbricate; leaves never as above; rock plants **193. Gesneriaceae**
38a. (36) Ovules 1–2 per loculus 39
38b. Ovules 3–numerous per loculus 41
39a. (38) Arching shrub with small purple flowers in clusters on last year's wood **189. Buddlejaceae**
39b. Combination of characters not as above 40
40a. (39) Sepals free; corolla lobes contorted and infolded, rarely imbricate; twiners, herbs or dwarf shrubs **177. Convolvulaceae**
40b. Sepals connate; corolla lobes imbricate; trees or shrubs **182. Boraginaceae**
41a. (38) Corolla lobes usually folded, contorted or valvate; septum often oblique; internal phloem present **176. Solanaceae**
41b. Corolla lobes imbricate; septum horizontal; internal phloem absent **190. Scrophulariaceae**
42a. (26) Trailing, heath-like shrublets **90. Ericaceae**
42b. Plants not as above 43
43a. (42) Milky sap usually present; fruit often of 2 'follicles' and seeds with silky appendages 44
43b. Milky sap absent; fruit a capsule or fleshy; seeds without silky appendages 45
44a. (43) Pollen granular, transferred directly from anthers; corona absent; corolla lobes contorted in bud **172. Apocynaceae**
44b. Pollen often in pollinia, transferred by means of specialised translators; corona usually present; corolla lobes contorted or valvate in bud **173. Asclepiadaceae**
45a. (43) Ovules 1–2; trees with 4-merous flowers in axillary panicles **134. Salvadoraceae**
45b. Ovules 4–many; herbs, shrubs or trees; flowers 4–5-merous 46
46a. (45) Flowers in scorpioid cymes; herbs **181. Hydrophyllaceae**
46b. Flowers not in scorpioid cymes; herbaceous or woody 47
47a. (46) Placentation parietal; carpels 2 48
47b. Placentation axile; carpels 2–3 49
48a. (47) Leaves compound; epicalyx present **181. Hydrophyllaceae**
48b. Leaves simple; epicalyx absent **171. Gentianaceae**
49a. (47) Stamens fewer than corolla lobes **184. Verbenaceae**
49b. Stamens as many as corolla lobes 50
50a. (49) Leaves exstipulate; carpels 3; corolla lobes contorted in bud; herbaceous **180. Polemoniaceae**

50b. Leaves stipulate (stipules often reduced to a ridge between the leaves); carpels usually 2; corolla lobes imbricate or valvate in bud; mostly woody 51

51a. (50) Corolla (4–)5-lobed; stellate and/or glandular hairs absent; intraxylary phloem present **170. Loganiaceae**

51b. Corolla 4-lobed; stellate and/or glandular hairs present; intraxylary phloem absent **189. Buddlejaceae**

## Group 11

*Calyx and corolla both present; petals all united, at least at the base; corolla zygomorphic; ovary superior.*[1]

1a. Stamens more numerous than the corolla lobes, or anthers opening by pores 2

1b. Stamens as many as corolla lobes or fewer, not opening by pores 6

2a. (1) Anthers opening by pores; leaves undivided; ovary syncarpous 3

2b. Anthers opening by slits; leaves dissected or compound; ovary of 1 carpel 5

3a. (2) Two lateral sepals petaloid; filaments united **166. Polygalaceae**

3b. No sepals petaloid; filaments free 4

4a. (3) Shrubs with alternate or whorled leaves; stamens 5–25 **90. Ericaceae**

4b. Herbs with opposite leaves; stamens 5 **171. Gentianaceae**

5a. (2) Leaves pinnate or trifoliolate; perianth not spurred (see p. 60) **108. Leguminosae** s.l.

5b. Leaves laciniate; perianth spurred **18. Ranunculaceae**

6a. (1) Stamens as many as corolla lobes; zygomorphy weak 7

6b. Stamens fewer than corolla lobes, 4 or 2; zygomorphy pronounced 12

7a. (6) Stamens antipetalous; placentation free-central **100. Primulaceae**

7b. Stamens antisepalous; placentation axile 8

8a. (7) Leaves with pellucid aromatic glands, trifoliolate; stamens 5, the 2 upper fertile, the 3 lower sterile **153. Rutaceae**

8b. Combination of characters not as above 9

[1] Many of the bicarpellate families in group 11 are rather weakly delimited; it is not possible to account for all the exceptions.

9a. (8) Ovary 3-carpellate; ovules many **180. Polemoniaceae**
9b. Ovary 2-carpellate; ovules 4 or many 10
10a. (9) Flowers in scorpioid cymes; fruit of 4 1-seeded nutlets **182. Boraginaceae**
10b. Flowers not in scorpioid cymes; fruit a many-seeded capsule 11
11a. (10) Corolla contorted; stamens 5, equal; leaves opposite, subdeciduous; climber **170. Loganiaceae**
11b. Corolla imbricate; stamens 4 or unequal; leaves usually alternate, usually deciduous **190. Scrophulariaceae**
12a. (6) Placentation axile; ovules 4 or many 13
12b. Placentation parietal, free-central, basal or apical; ovules many or 1–2 20
13a. (12) Ovules numerous but not superposed (i.e. not in a vertical row in each loculus) 14
13b. Ovules 4, or more numerous and superposed 16
14a. (13) Seeds winged; mainly trees, shrubs or climbers with opposite, pinnate, digitate or rarely simple leaves **195. Bignoniaceae**
14b. Seeds usually wingless; mainly herbs or shrubs with simple leaves 15
15a. (14) Corolla lobes variously imbricate in bud; septum horizontal; leaves opposite or alternate; internal phloem absent **190. Scrophulariaceae**
15b. Corolla lobes usually folded, contorted or valvate in bud; septum usually oblique; leaves alternate; internal phloem present **176. Solanaceae**
16a. (13) Leaves all alternate, often studded with resinous glands; woody **191. Myoporaceae**
16b. At least the lower leaves opposite or whorled; herbaceous or woody 17
17a. (16) Fruit a capsule; ovules 4–many, usually superposed 18
17b. Fruit not a capsule; ovules 4, side by side 19
18a. (17) Leaves all opposite, often prominently marked with cystoliths; pedicels without swollen glands at the base; capsule opening elastically, seeds usually on hooked funicles **196. Acanthaceae**
18b. Upper leaves alternate, without cystoliths; pedicels with swollen glands at the base; capsule not elastic, seeds not on hooked funicles **197. Pedaliaceae**
19a. (17) Style gynobasic, or if terminal then corolla with a reduced upper lip; fruit usually of 4 1-seeded nutlets; corolla often strongly bilabiate; calyx often 2-lipped **185. Labiatae**

19b. Style terminal; upper lip of corolla well-developed; fruit usually a berry or drupe; corolla often less strongly zygomorphic; calyx ± actinomorphic **184. Verbenaceae**

20a. (12) Ovules 4–many; fruit a capsule, rarely a berry or drupe 21

20b. Ovules 1–2; fruit indehiscent, often dispersed in the calyx 27

21a. (20) Ovules 4, side by side **184. Verbenaceae**

21b. Ovules many 22

22a. (21) Placentation free-central; corolla spurred **198. Lentibulariaceae**

22b. Placentation parietal; corolla not spurred, rarely saccate 23

23a. (22) Leaves scale-like, never green; root parasites 24

23b. Leaves green, expanded; free-living plants 25

24a. (23) Placentas 4; calyx laterally 2-lipped **194. Orobanchaceae**

24b. Placentas 2; calyx ± equally 4-lobed **190. Scrophulariaceae**

25a. (23) Seeds winged; mainly climbers with opposite, pinnately divided leaves **195. Bignoniaceae**

25b. Combination of characters not as above 26

26a. (25) Capsule with a long beak separating into 2 curved horns; plant glutinous-villous **197. Pedaliaceae**

26b. Capsule without beak or horns; plants often softly velvety or glabrous **193. Gesneriaceae**

27a. (20) Flowers in involucrate capitula; ovule 1, apical **192. Globulariaceae**

27b. Flowers not in involucrate capitula, often in spikes; ovules 1 or 2 28

28a. (27) Flowers deflexed in fruit; calyx with hooked teeth; ovary unilocular with 1 basal ovule **186. Phrymaceae**

28b. Flowers erect in fruit; calyx not hooked; ovary bilocular with the ovules solitary at the top of each loculus; fruit often 1-seeded **190. Scrophulariaceae**

## Group 12

*Calyx and corolla both present, the former sometimes reduced to a rim; petals all united, at least at the base; ovary partly or fully inferior.*

1a. Inflorescence an involucrate capitulum (or the flowers rarely in distant, spiny-bracted whorls); ovules always solitary 2

1b. Inflorescence and ovules not as above 3

2a. (1) Each flower with a cup-like involucel; stamens 4, free; ovule apical **207. Dipsacaceae**
2b. Involucel 0; stamens 5, syngenesious; ovule basal **208. Compositae**
3a. (1) Stamens 2; stamens and style united into a sensitive column; leaves linear **200. Stylidiaceae**
3b. Combination of characters not as above 4
4a. (3) Leaves alternate or all basal 5
4b. Leaves opposite or appearing whorled 13
5a. (4) Anthers opening by pores; fruit a berry or drupe **90. Ericaceae**
5b. Anthers opening by slits; fruit various 6
6a. (5) Tendrillar climbers with unisexual flowers; stamens 1–5; placentation parietal; fruit berry-like **81. Cucurbitaceae**
6b. Combination of characters not as above 7
7a. (6) Stamens 10–many; woody plants 8
7b. Stamens 4–5; mainly herbs 10
8a. (7) Leaves gland-dotted, smelling of eucalyptus; corolla, calyptrate **117. Myrtaceae**
8b. Combination of characters not as above 9
9a. (8) Indumentum of stellate or lepidote hairs; stamens 1-seriate; anthers linear **96. Styracaceae**
9b. Indumentum, when present, not as above; stamens several-seriate; anthers broad **97. Symplocaceae**
10a. (7) Stigma surrounded by a sheath **202. Goodeniaceae**
10b. Stigma not surrounded by a sheath 11
11a. (10) Stamens antipetalous **100. Primulaceae**
11b. Stamens not antipetalous 12
12a. (11) Stamens epipetalous, 2 or 4; milky sap absent **193. Gesneriaceae**
12b. Stamens ± free from corolla, 5 or rarely more; milky sap usually present **199. Campanulaceae**
13a. (4) Placentation parietal; stamens 2 or 4 and paired **193. Gesneriaceae**
13b. Placentation axile or apical; stamens 1 or more, not paired 14
14a. (13) Stamens 1–3; ovary with 1 ovule; fruit a cypsela **206. Valerianaceae**
14b. Stamens 4 or more; ovary usually with 2 or more ovules; fruit not a cypsela 15
15a. (14) Leaves ternately compound; flowers in a few-flowered head like a clock-tower; rhizomatous herb **205. Adoxaceae**

15b. Leaves simple or rarely pinnate; inflorescence usually cymose, sometimes spicate, capitular, or flowers solitary; woody or herbaceous 16

16a. (15) Stipules interpetiolar (rarely intrapetiolar), sometimes leaf-like (leaves thus appearing whorled); ovary usually 2-locular; flowers usually actinomorphic; fruit capsular, fleshy or schizocarpic **203. Rubiaceae**

6b. Stipules usually absent, when present not as above; ovary (2–)3(–5)-locular, sometimes only 1 loculus fertile; flowers often zygomorphic, sometimes twinned; fruit a berry or drupe **204. Caprifoliaceae**

## MONOCOTYLEDONS

### Group 13

*ary superior or naked; perianth well-developed, or reduced or absent; luding all aquatics with totally submerged flowers.*

a. Trees, shrubs or prickly scramblers with large, plicate, palmately or pinnately divided leaves; flowers ± sessile in fleshy spikes or panicles with spathe-like bracts **242. Palmae**

1b. Combination of characters not as above 2

2a. (1) Plant body thallus-like, not differentiated into stem and leaf; minute monoecious floating aquatics **246. Lemnaceae**

2b. Plant body differentiated into stem and leaf; terrestrial, aquatic or rarely epiphytic 3

3a. (2) Totally submerged aquatics of fresh or saline water 4

3b. Terrestrial or epiphytic plants, or if aquatic then not with the flowers submerged 9

4a. (3) Perianth of 4 clawed valvate segments; aquatics of fresh or brackish water with bisexual flowers in axillary spikes; carpels 4, free, sessile **217. Potamogetonaceae**

4b. Combination of characters not as above; perianth reduced or absent 5

5a. (4) Marine plants with densely fibrous rhizome (washed up on beaches as fibre balls); leaves mostly basal, ligulate; flowers borne in pedunculate spikes subtended by reduced leaves **221. Posidoniaceae**

5b. Combination of characters not as above 6

6a. (5) Flowers axillary, not enclosed by a leaf sheath; plants of fresh or brackish water 7

6b. Flowers on flattened axes or in 2-flowered spikes, at first enclosed by leaf sheaths; plants of brackish water, or marine 8

7a. (6) Leaves toothed; ovary solitary with 2–4 slender stigmas **216. Najadaceae**

7b. Leaves entire; carpels 1–9, free, stigma dilated or 2–4-lobed **219. Zanichelliaceae**

8a. (6) Flowers in 2-flowered spikes, long-peduncled in fruit; carpels free, 4 or more, becoming long-stalked in fruit to form a false umbel **218. Ruppiaceae**

8b. Flowers numerous on flattened axes lined at the margins with solitary bract-like perianth segments; ovary solitary; fruit subsessile **220. Zosteraceae**

9a. (3) Stemless, monoecious herbs; leaves plicate, fan-shaped or deeply bifid; flowers reduced, crowded on a pedunculate spadix subtended by caducous spathes **243. Cyclanthaceae**

9b. Combination of characters not as above 10

10a. (9) Flowers in involucrate capitula, unisexual, minute; perianth scarious or membranous, greyish white **225. Eriocaulaceae**

10b. Combination of characters not as above 11

11a. (10) Dioecious trees or shrubs with stiff leathery leaves, and often supported by stilt roots; fruit a syncarp, often woody **244. Pandanaceae**

11b. Combination of characters not as above 12

12a. (11) Inflorescence a simple fleshy spike (spadix) of inconspicuous flowers subtended by (rarely adnate to) a large bract (spathe); leaves often net-veined and/or lobed; rarely small floating aquatics **245. Araceae**

12b. Combination of characters not as above 13

13a. (12) Perianth entirely scarious, or reduced to bristles, hairs, narrow scales or lodicules, or absent 14

13b. Perianth well-developed, never scarious throughout, sometimes small or reduced to a single segment 19

14a. (13) Flowers imbricated in distichous or cylindrical spikelets (sometimes 1-flowered), each flower subtended by a membranous or tough bract 15

14b. Flowers arranged in heads, superposed spikes, panicles, cymes or clusters, but never imbricated into regular spikelets 17

15a. (14) Ovary with 2–3 ovules; flowers unisexual, plant usually dioecious; leaves reduced to sheaths **227. Restionaceae**

15b. Ovary with 1 ovule; flowers bisexual, or if unisexual then plants usually monoecious; leaves usually with well-developed blades (except in some *Cyperaceae* and *Juncaceae*) 16

16a. (15) Leaf phyllotaxis ½; stem usually cylindrical with hollow internodes; leaf sheath usually with free margins; flowers arranged in distichous spikelets (sometimes 1-flowered) subtended by 2 sterile bracts (glumes); each flower enclosed by an abaxial bract (lemma) and delicate adaxial bracteole (palea, rarely absent); perianth reduced to (3–)2(–0) lodicules **228. Gramineae**

16b. Leaf phyllotaxis ⅓; stem usually cylindrical or 3-sided with solid internodes; leaf sheath usually closed; flowers arranged in distichous or cylindrical spikelets or spikes usually not subtended by 2 sterile bracts; each flower subtended by 1 abaxial bract (glume), bracteole always absent; ovary sometimes enclosed in a flask-like structure; perianth of bristles, hairs, scales or 0 **230. Cyperaceae**

17a. (14) Flowers bisexual; perianth segments 6, scarious; ovules 3–many per ovary **229. Juncaceae**

17b. Flowers unisexual; perianth segments more reduced; ovules 1 per ovary 18

18a. (17) Flowers in 2 cylindrical superimposed brownish spikes; ovary on a hairy stalk **232. Typhaceae**

18b. Flowers in globose heads; ovary subsessile **231. Sparganiaceae**

19a. (13) Carpels free or rarely connate near base; aquatic or marsh plants 20

19b. Carpels solitary or ovary syncarpous though styles sometimes free; terrestrial, epiphytic, aquatic or marsh plants 25

20a. (19) Inflorescence spicate (sometimes interrupted or bifid); perianth segments 1–4 or 0 21

20b. Inflorescence never spicate, though sometimes a raceme; perianth segments 6 22

21a. (20) Stamens 6 or more; carpels 3–6; perianth segments 1–3, sometimes bract-like **213. Aponogetonaceae**

21b. Stamens, carpels and perianth 4-merous **217. Potamogetonaceae**

22a. (20) Carpels with diffuse-parietal placentation 23

22b. Carpels with basal or marginal placentation 24

23a. (22) Leaves linear, latex absent; perianth petaloid **209. Butomaceae**

23b. Leaves with petiole and blade, latex present; sepals and petals markedly different **210. Limnocharitaceae**

24a. (22) Leaf sheaths ligulate; flowers in racemes; perianth segments similar and all ± sepaloid **214. Scheuchzeriaceae**

24b. Leaf sheaths eligulate; flowers in whorls, racemes, panicles or umbels; perianth differentiated into sepals and petals **211. Alismataceae**

25a. (19) Perianth apparently 1-seriate, or the inner and outer whorls similar, often petaloid 26

25b. Perianth biseriate, with the outer and inner whorls manifestly different, the outer often calyx-like, the inner usually petaloid 31

26a. (25) Raceme or panicle subtended by an entire, spathe-like sheath; plants aquatic **247. Pontederiaceae**

26b. Inflorescence not as above; usually terrestrial or marsh plants 27

27a. (26) Plants scapose with small flowers in ebracteate racemes or spikes; perianth sepaloid; ovules 1 per cell, basal **215. Juncaginaceae**

27b. Plants with inflorescence not as above; perianth usually petaloid; ovules usually more than 1 per cell, rarely basal 28

28a. (27) Perianth persistent, covered with branched hairs; stamens 3, sap often orange **252. Haemodoraceae**

28b. Combination of characters not as above 29

29a. (28) Flowers 2-merous; ovary unilocular with apical or basal ovules; erect or climbing herbs **254. Stemonaceae**

29b. Flowers usually 3-merous (rarely 2-merous or more than 3-merous in *Liliaceae* s.l.); ovary with axile (rarely parietal) placentation; habit diverse 30

30a. (29) Usually climbing herbs bearing panicles; leaves distichous with closed sheaths and often ending in tendrils; perianth ± membranous; ovule 1 per loculus **226. Flagellariaceae**

30b. Usually erect herbs with rhizomes, corms or bulbs but sometimes either woody or climbing or with cladodes, succulent leaves, or annual; perianth petaloid; ovules usually many, rarely 1 per loculus (see p. 60) **248. Liliaceae** s.l.

31a. (25) Flowers solitary or umbellate 32

31b. Flowers clustered in spikes, heads, cymes or panicles 33

32a. (31) Aquatic herbs with submerged linear leaves 2-toothed at the apex **223. Mayacaceae**

32b. Terrestrial herbs with broad leaves whorled at the top of the stem, or opposite (see p. 60) **248. Liliaceae** s.l.

33a. (31) Outer perianth whorled with 1 segment much larger than the other 2; inner whorl petaloid, yellow; flowers in bracteate heads **222. Xyridaceae**

33b. Combination of characters not as above 34

34a. (33) Stamens 6 or 3, staminodes 0–3; anthers basifixed; leaves usually cauline, often with closed sheaths; bracts not conspicuous and coloured **224. Commelinaceae**

34b. Stamens 6, staminodes 0; anthers usually versatile; leaves mostly basal, often rigid and spiny-margined, rarely flaccid cauline and grey with peltate scales; bracts often conspicuous and coloured **233. Bromeliaceae**

## Group 14

*Ovary fully or partly inferior; perianth well-developed, usually petaloid; if aquatic, always with flowers at or above the water surface.*

1a. Flowers actinomorphic or sometimes weakly zygomorphic; stamens 6, 4, 3 or rarely many 2

1b. Flowers strongly zygomorphic, or asymmetrical; stamens (6–)5, 2 or 1 10

2a. (1) Unisexual climbers with cordate or digitate leaves; rootstock tuberous or woody **255. Dioscoreaceae**

2b. Combination of characters not as above 3

3a. (2) Perianth persistent, actinomorphic and villous, or weakly zygomorphic with branched hairs; sap often orange **252. Haemodoraceae**

3b. Combination of characters not as above 4

4a. (3) Rooted or floating aquatics; stamens 3–many; flowers unisexual or bisexual; placentation diffuse-parietal **212. Hydrocharitaceae**

4b. Usually terrestrial, epiphytic or marsh plants; stamens usually 3 or 6, rarely many; flowers usually bisexual, placentation axile or sometimes parietal 5

5a. (4) Stamens 3, staminodes 0; leaves often equitant; stylar branches often divided **249. Iridaceae**

5b. Stamens 6 (rarely 3+ 3 staminodes); leaves usually not equitant; styles not as above 6

6a. (5) Placentation parietal; inner bracts of scapose umbel often pendulous, long and thread-like **253. Taccaceae**

6b. Placentation usually axile; bracts not as above 7

7a. (6) Perianth whorls dimorphic with outer whorl ± calyx-like, the inner longer and petaloid; bracts often conspicuous and coloured; leaves usually stiff, mostly basal and often spiny-margined **233. Bromeliaceae**

7b. Perianth parts all similar and petaloid, sometimes weakly zygomorphic; leaves not as above 8

8a. (7) Anthers always poricidal; ovary half-inferior; flowers in racemes or panicles **250. Tecophilaeaceae**

8b. Anthers usually opening by longitudinal slits; ovary fully inferior; inflorescence diverse 9

9a. (8) Flowers in panicles, racemes, 'umbels' or solitary, usually bracteate; stamens always 6; herbs with rhizomes, corms, bulbs, or climbing, or plants with thick woody caudex, trunks or branches (see p. 60) **248. Liliaceae** s.l.

9b. Flowers always solitary, terminal on ebracteate peduncles rising from terminal tufts of leaves; stamens 6–many; 'dichotomously' branched shrubs with persistent leaf bases, or suffruticose herbs, or with flopping fibre-covered rhizomes **251. Velloziaceae**

10a. (1) Leaf veins (when visible) parallel to margin; median petal modified into a labellum; stamens 2 or 1, united with stigma-bearing column, pollen usually in sticky pollinia; ovary usually spirally twisted and placentation parietal; epiphytic or terrestrial **256. Orchidaceae**

10b. Leaf veins always spreading outwards from midrib; labellum, if present, often of staminodial origin; stamens (6–)5 or 1, pollen granular; ovary not spirally twisted, placentation usually axile; terrestrial 11

11a. (10) Fertile stamens (6–)5 12

11b. Fertile stamen 1, the remainder transformed into often petaloid staminodes, some of which form a labellum 15

12a. (11) Leaves and bracts spirally arranged; flowers unisexual; fruit a banana **237. Musaceae**

12b. Leaves and bracts distichous; flowers bisexual; fruit not a banana 13

13a. (12) Cymes arising from bases of leaf sheaths; sepals united below into a long stalk-like tube; median (abaxial) petal forming a large labellum **235. Lowiaceae**

13b. Flowers in a cincinnus in the axil of a spathe; sepals free or at most adnate to corolla; petals not forming a labellum 14

14a. (13) Perianth segments free; ovary with numerous ovules per loculus **234. Strelitziaceae**

14b. Perianth segments partially united; ovary with 1 ovule per loculus **236. Heliconiaceae**
15a. (11) Flowers strictly zygomorphic; sepals united into a tube; anther 2-celled, often supporting the style in a groove 16
15b. Flowers asymmetrical; sepals free; anther 1-celled, free from the style 17
16a. (15) Leaves distichous; ovary with prominent glands on top **238. Zingiberaceae**
16b. Leaves spirally arranged; ovary lacking epigynous glands **239. Costaceae**
17a. (15) Petiole not banded at junction with blade; ovary usually warty, with numerous ovules **240. Cannaceae**
17b. Petiole with a ± swollen band at junction with blade; ovary smooth, with 1–3 ovules **241. Marantaceae**

## KEYS TO SEGREGATE FAMILIES

### 106. Saxifragaceae s.l.

1a. Plants woody 2
1b. Plants herbaceous 5
2a. (1) Leaves apparently trifoliolate, sessile, opposite, evergreen (*Australia*) **Baueraceae**
2b. Leaves not as above 3
3a. (2) Stamens 8 or more; leaves usually opposite (*Mainly E Asia & N America*) **Hydrangeaceae**
3b. Stamens 4–5 (–6); leaves alternate 4
4a. (3) Disc present; leaves usually with gland-tipped teeth (*Mainly S America & Australasia*) **Escalloniaceae**
4b. Disc absent; leaves without gland-tipped teeth (*Temperate N Hemisphere, Andes*) **Grossulariaceae**
5a. (1) Stamens not alternating with staminodes (*Mainly North Temperate regions*) **Saxifragaceae** s.s.
5b. Stamens alternating with staminodes 6
6a. (5) Ovary globose; stamens 5; petals 5, staminodes multifid, gland-tipped (*North Temperate regions*) **Parnassiaceae**
6b. Ovary 4-sided; stamens 4 or 8; petals 4; staminodes simple (*Temperate S America*) **Francoaceae**

### 108. **Leguminosae/Fabaceae** s.l.

1a. Corolla actinomorphic; petals valvate; stamens 4–many; leaves bipinnate, rarely phyllodic; seeds with U-shaped lateral line (*Mainly Tropics and Subtropics; N America*) **Mimosaceae**

1b. Corolla zygomorphic (sometimes weakly so); petals imbricate, rarely 0; stamens 10 or fewer; leaves simply pinnate, trifoliate or simple; seeds usually without a lateral line, rarely with a closed line 2

2a. (1) Adaxial petal interior (ascending imbrication) or petal 1 or absent; seed usually with a straight radicle (*Mainly Tropical*) **Caesalpiniaceae**

2b. Adaxial petal exterior (descending imbrication); seed usually with an incurved radicle (*Widespread*) **Papilionaceae/Fabaceae** s.s.

### 248. **Liliaceae** s.l.

The traditional treatment of the *Liliaceae* complex, adopted in Bentham & Hooker's *Genera Plantarum* vol. 3, and still used in many Floras, distinguished 2 major families on ovary position: *Liliaceae* (ovary superior) and *Amaryllidaceae* (ovary inferior, including *Velloziaceae*, which are excluded here). The nine groups recognised on our segregate key could also be treated as subfamilies of *Liliaceae* s.l.

1a. Petiole bearing tendrils, or leaf surface reversed by a twisted petiole, or cladodes present in axils of reduced scale leaves 2

1b. Plants without any of the above characters 4

2a. (1) Ovary inferior; flowers large and showy; fruit a capsule (*C & S America*) **Alstroemeriaceae**

2b. Ovary superior; flowers small; fruit a berry 3

3a. (2) Plants with ovate, spinose or filiform cladodes; leaves scale-like or spinose (*Tropical & Temperate regions of the Old World*) **Asparagaceae**

3b. Shrubs without cladodes; leaves broad, with petioles bearing 2 tendrils (*Tropical & warm Temperate regions*) **Smilacaceae**

4a. (1) Herbs with 'umbellate' inflorescences, subtended by an involucre of membranous bracts, sometimes reduced to a single, bracteate, falsely terminal flower; perianth petaloid, with or without a corona; rootstock a bulb or rarely rhizome, leaves mostly basal; ovary superior or inferior (*Widespread*) **Amaryllidaceae**

4b. Combination of characters not as above 5

5a. (4) Caudex or woody trunk ending in a crown of large fibrous and/or fleshy leaves, or leaves sometimes petiolate and scattered on woody branches; flowers in panicles or racemes; ovary superior or inferior (excluding succulent-leaved plants with a narrowly tubular perianth and superior ovary) (*Mainly Tropical & Subtropical regions; N America*) **Agavaceae**

5b. Combination of characters not as above 6

6a. (5) Ovary inferior (Scapose herbs often with vertical corm-like stocks; leaves mostly basal, either hairy, prominently veined, or plicate; flowers in racemes, sub-umbels, heads or solitary, usually bracteate) (*S Hemisphere, Tropical Asia, Atlantic N America*) **Hypoxidaceae**

6b. Ovary superior 7

7a. (6) Perianth whorls markedly dissimilar, 3- or more-merous; leaves (except for basal scale leaves) opposite or usually in a whorl at the top of the stem (*N Temperate regions*) **Trilliaceae**

7b. Perianth whorls similar, petaloid, usually 3-merous; leaves not as above 8

8a. (7) Shrubs or woody climbers with scattered cauline leaves; flowers solitary, usually pendulous and large; placentation usually parietal; fruit a berry (*S Hemisphere*) **Philesiaceae**

8b. Usually herbs with rhizomes, corms or bulbs, rarely herbaceous or woody climbers; leaves basal (rarely reduced to sheaths), or cauline leaves spirally arranged or several-whorled, or if plant woody and/or with succulent leaves in terminal crowns then perianth tubular and sometimes inflated or zygomorphic; flowers never in umbels, 3-, rarely 2-merous; placentation usually axile; fruit a capsule or berry (*Widespread*) **Liliaceae** s.s.

# ARRANGEMENT AND DESCRIPTION OF FAMILIES

Considerations of space have allowed us to give only very short, telegraphic family descriptions. Reference should be made to p. x for a list of the abbreviations used. In general, the variation given in the descriptions is somewhat wider than that presented in the key, and many characters used in the latter have had to be omitted. Only the families keyed out in the main key are described here; the segregate families of the broad families (keyed out separately on pp. 59–61) are not described.

No attempt has been made to diagnose the orders in which the families are grouped, but for each we have attempted to cite a few differential features, most of which are repeated in the descriptions of the families. The circumscription and content of the orders remain largely matters of opinion, but they may serve to bring together those families which (on the whole) resemble one another most closely.

The families and orders have been listed within the ten superorders used in Stebbins's classification; these higher groups are differentiated on the basis of a number of characteristics, many of them microscopic or chemical, and merely serve to indicate broad relationships. Whether they represent phylogenetic relationships is another matter.

For various reasons, we have deviated from Stebbins's system in dealing with a small number of families; these are indicated in table 2.

In using the descriptions, the following features must be assumed for most species of a family, unless otherwise stated: milky sap absent, habit not succulent, parts of the flower free from each other, stamens not antipetalous, and anthers opening by slits.

The following points concerned with presentation must be noted:

*Morphology*

The oblique stroke (/) is used instead of 'or'; the letter 'n' is used instead of 'many' (i.e. more than 10 or 12). K – number of calyx segments, C – number of corolla segments, P – number of perianth segments when these are undifferentiated, A – number of stamens, G – number of

Table 2

| Stebbins's system | This book |
|---|---|
| *Betulaceae* | Divided into *Betulaceae* and *Corylaceae*. |
| *Molluginaceae* | Dismantled and the various parts included in *Aizoaceae*, *Phytolaccaceae* and *Caryophyllaceae*.[1] |
| *Pyrolaceae* | Included in *Ericaceae*. |
| *Monotropaceae* | Included in *Ericaceae*. |
| *Saxifragaceae*, *Hydrangeaceae* and *Grossulariaceae* | Included in *Saxifragaceae* s.l., in which seven segregate families are recognised: *Saxifragaceae* s.s., *Francoaceae*, *Parnassiaceae*, *Baueraceae*, *Hydrangeaceae*, *Grossulariaceae*, *Escalloniaceae*. |
| *Neuradaceae* | Included in *Rosaceae*. |
| *Leeaceae* | Included in *Vitaceae*. |
| *Apocynaceae* | Divided into *Apocynaceae* and *Asclepiadaceae*. |
| *Liliaceae*, *Agavaceae* and *Smilacaceae* | Included in *Liliaceae* s.l., in which nine segregate families are recognised: *Liliaceae* s.s., *Asparagaceae*, *Smilacaceae*, *Amaryllidaceae*, *Agavaceae*, *Hypoxidaceae*, *Trilliaceae*, *Philesiaceae*, *Alstroemeriaceae*. |
| *Zosteraceae* | Divided into *Zosteraceae* and *Posidoniaceae*. |

[1] The content and circumscription of the morphologically and chemically heterogeneous *Molluginaceae* are controversial. Our treatment does not conflict with the (admittedly limited) phytochemical information currently available to us.

carpels; these letters are also used in the collective sense, e.g. 'A antipet' means 'stamens antipetalous'. Brackets are used to indicate that the segments of any particular whorl are fused, e.g. C(5) means a corolla of 5 united segments.

In the dicotyledons we have usually not indicated whether the ovary is superior or inferior; we have, however, always stated whether the perianth and stamens are hypogynous, perigynous or epigynous, which is to give the same information (cf. pp. 4–12). Many of the gamopetalous dicotyledons are described as 'K hypog CA perig'; this indicates that the stamens are epipetalous and the ovary superior. In the monocotyledons the ovary position is always indicated.

In plants with an inferior ovary, the number of calyx segments is shown as free if the segments are completely free above their point of insertion on the ovary, and united if they are united above that point. This information is very difficult to obtain, and should be used with caution.

Information about inflorescence type is always given, but this, again, is difficult to obtain and condense; the information given here should not be regarded as a complete description of the range of inflorescence types found in any particular family. We have referred to six main inflorescence types: racemes, spikes, panicles, cymes, umbels and capitula; fascicles and clusters have also been used when the nature of the inflorescence is not easily understood. 'Racemose' has been used when spikes, racemes, or basically indeterminate panicles occur in the same family; 'cymose' has been used in a similar way to cover various types of determinate inflorescences.

Ovule number refers to the number of ovules in each free carpel (in apocarpous gynoecia) or syncarpous ovary, unless qualified by 'per cell'.

*Geography*
This is indicated in two ways: families with species native to Europe are marked with an asterisk (*); and those with species native to America north of 30° N are marked with a dagger (†). A short summary is also given of the overall distribution for each family. Most of the terms for this are self-explanatory; a few, however, need a little explanation. 'Trop' is given for families which occur in both the tropics and subtropics; families occurring *only* in the subtropics are cited as 'Subtrop'. 'Temp' includes the temperate regions of both hemispheres, unless qualified by N or S; in most cases Arctic and Antarctic zones are covered by 'Temp'. We have used America for the whole of the American continent and associated islands; Old World has been used for the eastern hemisphere.

For monogeneric families we have given the name of the genus in parentheses at the end of the description. Some synonyms, and short comments about the relationships of individual families, are also included.

The permitted alternative family names (for the eight families whose traditionally used names do not end in '-aceae') are given, separated from the more familiar name by a stroke, e.g. 169. *Umbelliferae/Apiaceae*.

## SUBCLASS DICOTYLEDONES

### Superorder Magnoliidae

Flowers often apocarpous, always polypetalous or apetalous, trimery frequent; stamens usually numerous and centripetal in development. Pollen binucleate and often 1-aperturate.

MAGNOLIALES

Aromatic trees/shrubs; pollen often 1-aperturate; G1–n and apocarpous. (*A broader view often includes Illiciales and Trochodendrales.*)

**1. Magnoliaceae.**† Trees/shrubs with vessels. Lvs simple with large deciduous stip. Fls bisex, act, solit. PA/KCA hypog, P/KC in several series, 3- or sometimes 4-merous, A n, spirally arranged, G n, spirally arranged; ov 2–n. Fr a group of follicles. Seeds large. *Mostly N Temp & Subtrop*.

**2. Winteraceae.** Woody, lacking vessels. Lvs alt, simple, exstip. Fls bisex, act, in cymes or fascicles. KCA hypog. K2–6/(2–6), valvate, C 2–several-seriate, A several, pollen 1-aperturate, G1–n in 1 whorl; ov 1–n, marginal. Follicle/berry. *Tropics (except Africa), S Temp.*

**3. Annonaceae.**† Woody. Lvs simple, exstip. Infl var. Fls usu bisex, act. KCA hypog. K usu 3, C3–6, A n, crowned by enlarged connective, G n, usu stalked in fr, rarely united into a mass; ov 1–n, basal/marginal. Seeds arillate, endosperm convoluted. *Trop.*

**4. Myristicaceae.** Dioecious trees. Lvs alt, entire, exstip. Infl racemose. Fls unisex, act. PA hypog. P(2–5) usu (3), valvate; male: A2–20, filaments united; female: G1; ov 1, basal. Fr fleshy, 2-valved. Seed with coloured aril, endosperm convoluted. *Trop.*

ILLICIALES

Woody; K & C numerous, poorly differentiated; pollen 3-aperturate; apocarpous.

**5. Illiciaceae.**† Woody, aromatic, with vessels. Lvs alt to whorled, simple, exstip. Fls bisex, act, solit/supra-axillary. KCA hypog. K & C7–n, imbricate, A4–n, pollen 3-aperturate, G5–n in 1 whorl; ov 1, sub-basal. Fr a group of follicles. *SE Asia, SE North America (Illicium).*

**6. Schizandraceae.**† Woody climbers. Lvs alt, simple, exstip. Fls unisex, act, axillary. KCA hypog. K & C9–15, poorly differentiated, A n, ± united into a fleshy mass, G n; ov 2–3. Fr baccate, crowded or distant on elongate axis. *N America, E Asia.*

LAURALES

Usu woody, aromatic; lvs exstip; KCA/PA usu perig/epig.

**7. Monimiaceae.** Trees/shrubs. Lvs opp, simple, exstip, usu evergreen. Fls solit/cymose, usu unisex, act. KCA/PA perig. K4–n, C4–n/P4–n, A usu n, G1–n; ov 1. Achenes in accrescent perig cup. *Mainly Trop.*

**8. Calycanthaceae.**† Shrubs. Lvs opp, entire, exstip. Fls solit, bisex, act. PA perig. P n, A5–30, G n; ov 1–2, marginal. Achenes. *SE USA.*

**9. Lauraceae.***† Woody. Lvs usu alt, entire, exstip, glandular-punctate. Infl cymose/racemose. Fls small, bisex/unisex, act. PA hypog/perig. P usu (6), A12 or variable, valvular, G1; ov 1, apical. Drupe-like berry. *Mostly Trop.*

**10. Chloranthaceae.** Herbs/woody. Lvs opp, simple, stip. Infl spike/panicle/capitulum. Fls unisex, apetalous, act; male: P0, A1–3, sometimes $\frac{1}{2}+1+\frac{1}{2}$, united; female: P3, epig, G1-loc; ov 1, pendulous. Drupe. *Trop, S Temp.*

PIPERALES

Fls inconspicuous, bracteate, often in spikes; pollen 1-aperturate.

**11. Saururaceae.**† Herbs. Lvs alt, simple, stip. Infl spike/raceme. Fls bisex, act. P0, A6–8, G3–4/(3–4), sup/inf; ov 1–10 per cell, par/ax. Follicle/fleshy capsule. *Scattered.*

**12. Piperaceae.** Herbs/shrubs. Lvs usu alt, entire, stip. Infl spike. Fls minute, usu bisex, often sunk in spike. P0, A1–10, G(2–5), sup; ov 1, basal. Small drupe. *Trop.*

ARISTOLOCHIALES

Herbaceous or shrubby, often climbing; lvs alt, exstip; ov ax; capsule. Flowers or leaves bizarre.

**13. Aristolochiaceae.***† Herbs/climbers. Lvs alt, simple, often cordate, exstip. Infl var. Fls bisex, act/zyg. PA epig. P(3) often bizarre and foetid, A6–n often adnate to style, G(4–6); ov n, ax. Capsule. *Mostly Trop.*

**14. Nepenthaceae.** Herbs/shrubby climbers. Lvs alt, blades prolonged into insectivorous pitchers. Infl raceme/panicle. Fls unisex, act. PA hypog. P2–4, A4–24, G(3–4); ov n, ax. *Mainly Malaysia* (*Nepenthes – often placed in Sarraceniales*).

NYMPHAEALES

Aquatic herbs lacking vessels; P6–n, free; placentation usu marginal.

**15. Ceratophyllaceae.***† Submerged aquatics. Lvs whorled, much divided, exstip. Fls solit, unisex. PA hypog. P(10–15), A10–20, G1; ov 1, marginal. Nut. *Widespread* (*Ceratophyllum*).

**16. Nymphaeaceae.***† Rhizomatous aquatics. Lvs alt, cordate/peltate. Fls bisex, act, solit. KCA hypog or serially adnate to ovary. K3–6, C3–n, A n, G3–n/(3–n); ov few–n on walls of carpels. *Widespread.*

**17. Nelumbonaceae.**†Rhizomatous aquatics. Lvs alt, peltate. Fls bisex, alt, solit. PA hypog. P n, spirally arranged, coloured, A n, G n, sunk in pitted top of obconical torus enlarging in fr; ov 1, pendulous. *N America, S Asia, Australia* (*Nelumbo*).

RANUNCULALES

Lvs alt, often divided; A3–n; G1–n, usu apocarpous.

**18. Ranunculaceae.***† Usu herbs/climbers. Lvs usu alt, simple/compound, usu exstip. Infl var. Fls bisex, act/zyg. KCA/PA hypog. P4–n/K3–5, C2–n rar (4), A n, G1–n rar united; ov 1–n, marginal/basal/apical/rar axile. Achenes/follicles. *Mainly Temp.*

**19. Berberidaceae.***† Herbs/shrubs. Lvs alt, simple/divided, usu exstip. Infl cymose/racemose/fls solit. Fls bisex, act. KCA hypog. K4–6, C4–6 rar 9, A4–18 often antipet, G apparently 1; ov few, basal/marginal. Capsule/berry. *Mainly N Temp.*

**20. Lardizabalaceae.** Woody, usu climbers. Lvs alt, compound, exstip. Infl raceme-like. Fls usu unisex, act. PA hypog. P3–6/K3–6, C6, A6/(6), G3–15; ov n, marginal. Berries. *Scattered.*

**21. Menispermaceae.**† Usu woody climbers. Lvs alt, simple, exstip. Fls unisex, act. PA hypog. P6–merous, A3–n, usu 6, G3–6; ov 1, marginal. Drupe/achene. *Mostly Trop.*

PAPAVERALES

Polypetalous; fls act (often with 2 planes of symmetry)/zyg; ov par; seeds often arillate. (*See note under Capparales, p. 74.*)

**22. Papaveraceae.***†Herbs rar shrubs, with milky or coloured sap. Lvs alt/basal, rar opp, simple/divided, exstip. Infl cymose/solit. KCA hypog-/rar perig. K2–4–n, free/rar united (calyptrate), caducous, C0–n, usu 4, A n rar 4, G(2–n), rar free or almost so; ov 1–n, par. Capsule. *Mostly N Temp.*

**23. Fumariaceae.***† Herbs with colourless sap. Lvs alt/basal, usu divided, exstip. Infl racemose, rar cymose. KCA hypog. K2, caducous, C4 with 2 planes of symmetry/zyg, petals di- or trimorphic, 1 or 2 usu saccate, usu free at base, rar variously united, A4, free/united in 2 bundles each of $\frac{1}{2}+1+\frac{1}{2}$, G(2); ov n–1, par/basal. Capsule/nut. *Mostly N Temp.*

SARRACENIALES

Lvs forming insectivorous pitchers; style umbrella-shaped.

**24. Sarraceniaceae.**† Herbs. Lvs basal tubular pitchers. Fls solit/racemose, bisex, act. KCA/PA hypog. K4–5, C5–0, A n, G(3–5), style peltately dilated; ov n, ax. Capsule. *America.* (*Often placed with Droseraceae and Nepenthaceae.*)

## Superorder Hamamelidae

Flowers ± strongly reduced and often unisexual; perianth poorly developed or lacking; inflorescence often a catkin; placentation never parietal with numerous seeds; pollen never 1-aperturate. Plants always woody.

### TROCHODENDRALES

Trees/shrubs, lacking vessels; carpels weakly united; ov n, marginal.

**25. Tetracentraceae.** Woody. Lvs alt, simple, exstip. Infl catkin-like. Fls bisex, act. PA hypog. P4, A4, G(4); ov n, marginal. Capsule. *China, Burma* (*Tetracentron*).

**26. Trochodendraceae.** Woody. Lvs whorled, simple, exstip. Fls act, bisex, racemose/fasciculate. PA hypog. P minute or 0, A n, G(6–10); ov n, marginal. Fr dehiscent. *Japan* (*Trochodendron*).

### HAMAMELIDALES

Woody; G apocarpous to inferior; ov marginal/axile. (*27 & 28 often put with Magnoliales.*)

**27. Cercidiphyllaceae.** Trees. Lvs deciduous, simple, opp/alt, stip. Fls unisex, plants dioecious, axillary; male: subsessile, P4, A15–20; female: pedicillate, P4, hypog, G4–6; ov n, marginal. Follicles. *China, Japan* (*Cercidiphyllum*).

**28. Eupteleaceae.** Woody. Lvs alt, exstip. Fls act, bisex. P0, A n, G6–n, stalked; ov 1–3, marginal. Fr a group of stalked samaras. *Himalayas, Japan, China* (*Euptelea*).

**29. Platanaceae.***† Trees with exfoliating bark. Lvs alt, lobed, stip. Infl globose heads. Fls unisex, act. PA perig/hypog. P3–5/(3–5), A3–7, G5–9; ov 1(–2), marginal. Carpels cohering in fr. *N Temp* (*Platanus*).

**30. Hamamelidaceae.**† Woody, often with stellate hairs. Lvs alt, simple/lobed, stip. Infl var. Fls uni/bisex, act/zyg. KCA perig/epig. K(4–5), C0–4–5, A2–8, G(2); ov 1–2, ax. Woody capsule. *Scattered.*

### EUCOMMIALES

Trees; fls unisex, solit; fr a samara. (*Usually placed with Urticales.*)

**31. Eucommiaceae.** Trees with milky sap. Lvs alt, simple, exstip. Fls solit, unisex, act. P0, A4–10, G(2), naked; ov 1, apical. Samara. *China* (*Eucommia*).

### LEITNERIALES

Dioecious shrubs, catkinate; fr drupe.

**32. Leitneriaceae.**† Woody. Lvs alt, entire, exstip. Infl catkin. Fls variously interpreted; male: P0, A3–12; female: P3–8, G1–loc, sup; ov 1, lateral. Drupe. *USA* (*Leitneria*).

MYRICALES

Woody, aromatic, catkinate; fr drupe.
**33. Myricaceae.***† Woody. Lvs alt, entire/divided, exstip, aromatic, gland-dotted. Infl catkin. Fls usu unisex. P0, A2–20, usu 4–8, G(2), naked; ov 1, basal. Drupe. *N Hemisphere.*

FAGALES

Woody, monoecious; lvs stip; at least male infl usu catkin; G inf/naked; fr nut.
**34. Fagaceae.***† Woody. Lvs usu alt, simple, stip. Male infl often a catkin. Fls unisex; male: P(4–7), A4–n; female: P(4–7), G(3–6), inf; ov 2 per cell, ax. Nut enveloped in cupule. *Temp & Trop.*
**35. Betulaceae.***† Woody. Lvs alt, simple, stip. Infl catkin. Fls unisex; male: P(4), A2/4; female: P0, G(2), naked; ov 1 per cell, ax. Small, often winged nut in 'cones'. *Mainly N Temp.*
**36. Corylaceae.***† Woody. Lvs alt, simple, stip. Male infl catkin. Fls unisex; male: P0, A2–20; female: P lobed, G(2), inf; ov 1–2, ± apical. Nut clasped in a bract/cupule. *N Temp.*

CASUARINALES

Woody, monoecious; branches whip-like, lvs reduced.
**37. Casuarinaceae.** Woody, branches jointed. Lvs whorled, scale-like. Infl catkin. Fls unisex P0 (?), A1, G(2), naked; ov 2, par. Samaras in woody cones. *Australasia to Malaysia* (*Casuarina*).

## Superorder Caryophyllidae

Flowers usually polypetalous or apetalous; stamens, when numerous, developing centrifugally; pollen 3-nucleate, never 1-aperturate; placentation usually free-central or basal; seeds often with perisperm and a curved embryo; plants often containing betalains instead of anthocyanins.

CARYOPHYLLALES

C free/0; embryo usu strongly curved round perisperm; betalains usu present.
**38. Phytolaccaceae.**† Herbs/woody. Lvs alt, entire, exstip. Infl var. Fls usu bisex, act. PA hypog. P4–5/(4–5), A3–n, G1–(n), rar free; ov 1 or 1

per cell, basal/ax. Fr often fleshy. *Mainly American Trop & S Hemisphere.*
**39. Nyctaginaceae.**† Herbs/woody. Lvs usu opp, entire, exstip. Infl cymose. Fls usu bisex, act. PA hypog. P(5), petaloid, tubular, A(1–30), G1–loc; ov 1, basal. Achene, often in persistent P. *Trop.*
**40. Cactaceae.**† Mostly spiny stem-succulents. Lvs usu absent. Fls usu solit & bisex, act. KCA epig. K n, C n/(n), A n, G(3–n); ov n, par. Berry. *Mostly America.*
**41. Aizoaceae.***† Herbs/shrubs, usu leaf succulents. Lvs usu opp, simple, stip/exstip. Fls bisex, act. KCA hypog/perig/epig. K(5–8), C n, A n, G(3–5); ov usu n, ax/par. Fr usu capsule. *Mostly S Africa. (Incl most of Molluginaceae.)*
**42. Caryophyllaceae.***† Herbs rar shrublets. Lvs usu opp, entire, stip/exstip. Infl cymose/fls solit. Fls usu bisex, act. KCA hypog/perig, rar PA hypog/perig. K4–5/(4–5), C4–5/0, A3–10, G(2–5); ov usu n, free-central or 1 and basal. Capsule/nut rar fleshy. *Mostly N Temp.*
**43. Portulacaceae.***† Herbs/shrubs, often fleshy. Lvs alt/opp, simple, stip. Infl var. Fls bisex, act. KCA hypog/epig. K2/(2), rar 3, C3–12/(3–12) at base, usu 4–6, A3–n antipet when few, G(2–3); ov 1–n, basal/free central. Capsule. *Mostly New World.*
**44. Basellaceae.**† Climbers. Lvs alt, simple, exstip. Infl racemose. Fls bisex, act. PA perig. P5/(5), A5, G(3); ov 1, basal. Fr in persistent, fleshy P. *Mostly Trop America.*
**45. Chenopodiaceae.***† Herbs/shrubs. Lvs alt/opp, simple, exstip/reduced to scales when stem fleshy and segmented. Infl usu cymose. Fls uni/bisex, act. PA usu hypog. P(3–5) rar 0, green/membranous, A usu 5, G(2–3) rar ½–inf; ov 1, basal. Achene/nut. *Widespread.*
**46. Amaranthaceae.***† Herbs/woody. Lvs alt/opp, usu entire, exstip. Infl often racemose. Fls usu bisex, act. PA hypog. P3–5/(3–5), usu scarious, A5/(5), G(2–3); ov 1–few, basal. Capsule/achene. *Mostly Trop.*

BATALES

Dioecious shrubs; fr syncarp of berries.
**47. Bataceae.**† Shrubs. Lvs opp, simple, exstip. Infl catkin. Fls unisex: male: P2, A4–5 + 4–5 staminodes; female: P0, G(4), naked; ov 1 per cell, ascending. Syncarp of berries. *Coasts of America (Batis).*

POLYGONALES

Stip often prominent and united into a sheath; ov 1, basal; betalains absent.

**48. Polygonaceae.***† Herbs/woody. Lvs usu alt, simple, stip often united into a sheath (ochrea), rar reduced to a line. Infl often cymose. Fls usu bisex, act. PA usu hypog. P3–6, A6–9, G(2–4) usu (3); ov 1, basal. Nut. *Mostly N Temp.*

PLUMBAGINALES

A antipet; ov 1 basal, usu on long curved funicle; betalains absent.
**49. Plumbaginaceae.***† Herbs/shrubs. Lvs alt/basal, simple, exstip. Infl cymose/racemose/capitular. Fls bisex, act. KCA hypog/CA perig. K(5), C(5) rar 5, A5 antipet, G(5), stigmas 5; ov 1, basal. Fr indehiscent, retained in K tube. *Widespread. (Often assigned to Primulales.)*

## Superorder Dilleniidae

Flowers usually actinomorphic, polypetalous, gamopetalous or apetalous; stamens, when numerous, developing centrifugally; pollen usually binucleate; ovary sometimes apocarpous; placentation seldom free-central or basal; seeds without perisperm.

DILLENIALES

Usu shrubs; A n, centrifugal; fr usu follicular, seeds usu arillate.
**50. Dilleniaceae.** Woody, some climbing. Lvs alt, simple, stip/exstip. Infl var. Fls uni/bisex, act. KCA hypog. K5, C5, A n, often in bundles, G1–n; ov 1–n, marginal. Follicles/berry-like. *Australasia, Trop.*
**51. Paeoniaceae.***† Herbs/shrubs. Lvs alt, ternately compound, exstip. Fls usu solit, bisex, act. KCA hypog. K usu 5, heteromorphic, C5 or more, A n, G2–8; ov n, marginal. Large follicles. *N Temp* (*Paeonia*).
**52. Crossosomataceae.**† Shrubs. Lvs alt, simple. Fls solit, terminal, bisex, act. KCA perig. K5, C5, A15–n, G3–5; ov n, marginal. Follicles, seeds with multifid arils. *Western N America* (*Crossosoma*).

THEALES

Usu woody, often evergreen; G sup; ov usu ax.
**53. Ochnaceae.** Usu woody. Lvs alt, usu simple, stip. Infl var. Fls bisex, act. KCA hypog. K4–5/(4–5), C4–5, A5–n, poricidal, G3–15, united by a common style; ov 1–n per cell, ax. Usu schizocarp, often fleshy. *Mostly Trop.*
**54. Theaceae.**† Woody. Lvs alt, simple, exstip. Fls solit/fasciculate, usu bisex, act. KCA hypog/CA perig. K5, C5, sometimes adnate to A, A n–15, free or variously connate, G(3–5); ov n, ax. *Trop & warm Temp.*
**55. Stachyuraceae.** Woody. Lvs alt, simple, stip. Infl racemose. Fls usu

bisex, act. KCA hypog. K4, C4, A8, G(4); ov n, ax. Berry. *E Asia* (*Stachyurus*).

**56. Marcgraviaceae.** Woody, often epiphytic. Lvs alt, simple, exstip. Infl raceme/umbel; bracts of sterile fls variously modified into pitcher-like, pouched or spurred nectaries. KCA hypog. K4–5, C4–5/(4–5), calyptrate when united, A3–n, G(2–n); ov n, par. Capsule/indehiscent. *Trop America.*

**57. Elatinaceae.***† Small usu aquatic herbs. Lvs opp, simple, stip. Fls solit/cymose, bisex, act/zyg. KCA hypog. K3–5/(3–5), C3–5, A6–10, G(3–5); ov n, ax. Capsule. *Widespread.* (*Affinities uncertain.*)

**58. Guttiferae/Clusiaceae.***† Herbs/woody. Lvs opp, simple, usu exstip, gland-dotted. Infl cymose. Fls uni/bisex, act. KCA hypog. K2–10/(4–5), C3–12, A usu n, fascicled, G(3–5); ov 1–n per cell, ax/par. Capsule/berry. *Widespread.*

MALVALES

Stellate hairs common; K valvate; filaments often connate; G sup; ov ax.

**59. Elaeocarpaceae.** Woody. Lvs alt/opp, simple, stip. Infl var. Fls usu bisex, act. KCA hypog. K4–5/(4–5), C4–5, often laciniate, A usu n, poricidal, G(2–n); ov n, ax. Capsule/drupaceous. *Trop.*

**60. Tiliaceae.***† Usu woody, often with stellate hairs. Lvs usu alt, simple, stip. Infl cyme. Fls usu bisex, act. KCA hypog. K3–5 usu 5/(5), C0–5 usu 5, A10–n/(10–n), G(4–5); ov 1–n per cell, ax. Fr var. *Widespread.*

**61. Sterculiaceae.**† Woody, often with stellate hairs. Lvs alt, simple/divided, stip. Infl var. Fls usu bisex & act. KCA hypog. K(3–5), C0–5, A(5–10) rar free, G(4–5); 2–n per cell, ax. Fr var. *Mostly Trop.*

**62. Bombacaceae.** Trees, often with swollen trunk. Lvs simple/digitate, often lepidote, stip deciduous. Fls large, bisex, act. KCA hypog/CA perig. K5/(5), C5, crumpled in bud, A5–n/(5–n), anthers 1-celled, pollen smooth, G(2–5); ov 2–n, ax. Capsule/indehiscent, seeds often embedded in wool of pericarp. *Trop.*

**63. Malvaceae.***† Herbs/woody, often with stellate hairs. Lvs alt, simple/divided, stip. Infl var. Fls usu bisex, act. K hypog CA perig. K5/(5), often with epicalyx, C5, contorted, free but united with base of staminal tube, A (n) monadelphous, anthers 1-celled, pollen spiny, G(2–n); ov 1–n per cell, ax. Capsule/schizocarp. *Widespread.*

URTICALES

Lvs simple, alt; fls small, often unisex, apetalous; ovary usu sup; ov 1.

**64. Ulmaceae.***† Woody. Lvs alt, simple, stip, usu with oblique base.

Infl var. Fls uni/bisex, zyg. PA hypog. P(4–8), A4–8, G(2); ov 1, apical. Samara/drupe. *N Hemisphere.*

**65. Moraceae.**† Usu woody with milky sap. Lvs alt, simple, stip. Infl var, fls often sunk in expanded receptacle. Fls unisex, act, PA hypog. P2–6 usu 4, A1–4, G(2), normally 1 aborting; ov 1, apical. Syncarps frequent. *Mostly Trop.*

**66. Cannabaceae.***† Herbs/climbers. Lvs alt/opp, simple/divided, stip. Infl var. Fls unisex. PA hypog. Male: P(5), A5; female: P entire, enveloping ovary, G(2); ov 1, apical. Achene. *N Temp.*

**67. Urticaceae.***† Usu herbs, often with rough or stinging hairs. Lvs alt/opp, simple, usu stip. Infl var. Fls unisex, act. PA hypog/epig. P0–5/(2–5), A3–5 usu 4, inflexed in bud and sensitive, G1; ov 1, basal. Achene/drupe. *Widespread.*

VIOLALES

Woody/herbaceous; C usu free; G sup/inf; ov usu par.

**68. Flacourtiaceae.** Woody. Lvs alt, simple, stip. Infl var. Fls usu bisex, act. KCA/PA hypog/epig. K2–15, C0 or 2–15, A n, rar 5 or in antipet bundles, G(2–n); ov n, par. Capsule/berry. *Mostly Trop.*

**69. Violaceae.***† Herbs/shrubs. Lvs usu alt, simple/divided, stip. Infl var. Fls bisex, often zyg. KCA hypog. K5, C5, A5, G(3) rar (5); ov 1–n per cell, par. Capsule/berry. *Widespread.*

**70. Turneraceae.** Shrubs/herbs. Lvs alt, simple, exstip. Fls solit/racemose, bisex, act. KCA perig. K(5), C5, contorted, A5, G(3), stigmas brush-like; ov n, par. Capsule. *Mainly Trop America.*

**71. Passifloraceae.**† Often tendrillar climbers. Lvs alt, simple/compound, stip. Fls axillary, bisex, act. KC adnate below, A & G often on androgynophore. K4–5/(4–5), C4–5 rar 0, often with corona, A5/(5), G(3–5); ov n, par. Berry/capsule. *Mainly Trop America.*

**72. Bixaceae.**† Trees/shrubs. Lvs alt, simple, palmately nerved/lobed, stip. Infl panicle/raceme. KCA hypog. K5, imbricate, C5, imbricate, A n, G(2–5); ov n, par. 2–5-valved capsule. *Trop.*

**73. Cistaceae.***† Herbs/shrubs. Lvs usu opp, simple, stip/exstip. Infl usu cymose/fls solit/raceme. Fls bisex, act. KCA hypog. K3–5 often heteromorphic, C3–5 rar 0, A n, G(3–10) usu (5); ov 2–n per cell, par. Capsule. *Mainly warm N Temp.*

**74. Tamaricaceae.***† Woody. Lvs alt, usu scale-like, exstip. Infl usu raceme. Fls bisex, act. KCA hypog. K4–5, C4–5, A4–10 often arising from a disc, G(3–4); ov n, par/basal. Capsule, seeds bearded. *Mostly Mediterranean & C Asia.*

**75. Frankeniaceae.***† Halophytic shrubs/herbs. Lvs opp, entire, exstip.

Infl cymes/fls solit. Fls bisex, act. KCA hypog. K(4–7), C4–7, ligulate, A4–7 usu 6, free or united, G(3); ov n, par. Capsule. *Widespread.*
**76. Fouquieriaceae.**† Woody, spiny. Lvs alt, simple, exstip, fleshy. Infl terminal panicle. Fls bisex, act. K hypog CA perig. K5, C(5), A10–17, G(3); ov 12–18, par. Capsule. *C & SE North America.*
**77. Caricaceae.** Woody. Lvs alt, long-petioled, divided, exstip. Infl var. Fls heteromorphic, often unisex. KCA hypog. K(5), C5/(5), A10–5, G(5); ov n, par. Berry. *Trop America & Africa.*
**78. Loasaceae.**† Herbs/shrubs often with rough/stinging hairs. Lvs opp/alt, simple/divided, exstip. Fls axillary, bisex, act. KCA usu epig. K4–5, C4–5, A n connate in antipet bundles, G(3–7); ov n, ax/par. Capsule. *Mainly America.* (*Position uncertain.*)
**79. Begoniaceae.** Herbs/shrubs. Lvs alt, simple, stip, often fleshy, base often oblique. Fls usu unisex, act/zyg. KCA/PA epig. P2–12/K2, C2, A n/(n), G(2–5) usu (3); ov n, ax. Capsule/berry. *Mostly Trop.*
**80. Datiscaceae.**† Herbs/woody. Lvs alt, simple/compound, exstip. Infl var, Fls usu unisex, act, PA epig. K3–8/(3–8), C0–8, A4–n, G(3); ov n, par. Capsule. *Scattered.*
**81. Cucurbitaceae.***† Mostly herbs, tendrillar. Lvs alt, often lobed, exstip. Infl axillary cymes/fls solit. Fls usu unisex, act. KCA epig. K5/(5), C5/(5), A1–5 usu 3, 1 anther 1-celled, G(3–5); ov n, par, rar ax. Fr berry-like. *Widespread but mainly Trop.* (*Position uncertain.*)

### SALICALES

Dioecious, woody; ov par; fr capsule; seeds woolly.
**82. Salicaceae.***† Woody, dioecious. Lvs usu alt, simple, stip. Infl catkin. Fls usu unisex with disc or nectary gland. A2–n/(2–n), G(2–4), sup; ov n, par. Capsule. Seeds woolly. *Widespread.*

### CAPPARALES

Herbs/shrubs containing myrosin cells (smelling of mustard oils when crushed); C free; ov par; fr usu capsule (*A group formerly closely associated with Papaverales (p. 67), which they resemble closely, differing mainly in chemical and micromorphological characteristics.*)
**83. Capparaceae.***† Woody/herbs. Lvs alt, simple/compound, stip/exstip. Infl raceme/fls solit. Fls uni/bisex, act/zyg. KCA hypog. K4–8, C4–n, A4–n, G(2–4), gynophore frequent; ov few–n, par. Capsule/berry/nut. *Trop, warm Temp.*
**84. Cruciferae/Brassicaceae.***† Usu herbs. Lvs usu alt, simple/divided, exstip. Infl usu ebracteate raceme. Fls usu bisex & act (2 planes of symmetry). KCA usu hypog. K4, C4, A6 usu 2 + 4, rar 4 or n, G(2) rar

(4); ov 2–n, par. Usu capsule with false septum (replum). *Temp.*
**85. Resedaceae.***† Herbs/shrubs. Lvs alt, simple/divided, stip minute. Infl spike/raceme. Fls uni/bisex, zyg. KCA usu hypog. K4–8, C4–8 rar 2, A3–n, G(2–6) rar free; ov n, par. Capsule/berry, usu gaping at apex. *N Temp.*
**86. Moringaceae.** Trees. Lvs alt, 2–3-pinnate, exstip. Infl panicle. Fls bisex, zyg. KCA perig. K5, C5, A5 + 3–5 staminodes, anthers 1-celled, G(3); ov n, par. Triquetrous capsule. *Mostly Trop Old World* (*Moringa*).

ERICALES

Usu woody; lvs simple, exstip, often evergreen; A often poricidal.
**87. Actinidiaceae.** Woody. Lvs alt, simple, exstip. Infl cymes/fascicles/panicles. Fls uni/bisex, act. KCA hypog. K5, C5/(5), A10–n, often poricidal, G(3–5); ov n, ax. Berry/capsule. *Tropics.*
**88. Cyrillaceae.**† Woody. Lvs alt, simple, exstip. Infl raceme. Fls bisex, act. KCA hypog, disc 0. K(5), C5/(5), A10/5, G(2–4); ov 1–2 per cell, ax. Fr dry, indehiscent. *Mainly Trop America.*
**89. Clethraceae.**† Shrubs/trees. Lvs alt, simple, exstip. Infl raceme/panicle. Fls bisex, act, disc 0. KCA hypog. K(5), C5, A10–12, poricidal, pollen not in tetrads, G(3); ov n, ax. Capsule. *Mostly Trop & Subtrop* (*Clethra*).
**90. Ericaceae.***† Woody/herbs, rar saprophytic lacking chlorophyll. Lvs alt/opp/basal, simple, exstip. Infl racemose/clustered/fls solit. Fls bisex, usu act. KCA hypog/epig rar K hypog CA perig, disc present. K(4–5) rar 4–5, C(3–5) rar (10) rar free, A5–10 rar up to 25, usu poricidal, G(2–12); ov n, ax rar par. Capsule/berry/drupe. *Widespread.* (*Incl Pyrolaceae*, *Monotropaceae*, *Vacciniaceae.*)
**91. Epacridaceae.** Shrubs. Lvs alt, simple, exstip. Infl racemose. Fls bisex, act. K hypog CA perig. K(4–5), C(4–5), A4–5 anthers 1-celled, G1–10-loc; ov 1–n per cell, ax. Capsule/drupe. *Mostly Australasia.*
**92. Empetraceae.***† Shrublets. Lvs alt, entire, exstip. Infl var. Fls usu bisex, act. PA hypog, disc 0. P2–6 rar 0, A2–4, G(2–9); ov 1 per cell, ax. Drupe with 2–9 stones. *Temp.*

DIAPENSIALES

Shrublets; 5 staminodes present; A opening by slits.
**93. Diapensiaceae.***† Usu evergreen shrublets. Lvs alt, simple, exstip. Infl racemose/capitate/fls solit. Fls bisex, act. K hypog CA perig. K5/(5), C(5), A5 + 5–0 staminodes, G(3); ov usu n, ax. Capsule. *N Temp.*

EBENALES

Woody; lvs simple; fls act; A usu more numerous than C.

**94. Sapotaceae.**† Woody with milky sap. Lvs alt, simple, exstip, leathery. Infl cymose/fls solit. Fls bisex, act. K hypog CA perig. K(4–8), C(4–8), A antipet/2–3-seriate, staminodes frequent, G1–14-loc; ov 1 per cell, ax. Usu hard berry. *Mostly Trop.*

**95. Ebenaceae.**† Woody. Lvs alt, simple, entire, leathery. Infl cymose/fls solit. Fls usu unisex, act. KCA hypog/K hypog CA perig. K(3–7), C(3–7), A1–4 × as many as C, G(3–16); ov 1–2 per cell, ax. Berry. *Mostly Trop.*

**96. Styracaceae.***† Woody. Lvs alt, simple, exstip, with stellate/lepidote hairs. Infl var. Fls bisex, act. KCA hypog/epig/K hypog CA perig. K4–5/(4–5), C(4–5), A8–12, 1-seriate, G(3–5); ov n, ax at least below. Drupe, dehiscing irregularly. *America, Mediterranean, E Asia.*

**97. Symplocaceae.** Woody. Lvs alt, simple, exstip, leathery. Infl racemose/fls solit. KCA ± epig. K(5), C(3–11), A4–n, 1-4-seriate, G(2–5); ov 2–4 per cell, ax. Berry/drupe. *Warmer Asia, America, Australia* (*Symplocos*).

PRIMULALES

Fls usu act & gamopetalous; A antipet; ov free-central/basal.

**98. Theophrastaceae.** Woody. Lvs alt in false terminal whorls, simple, exstip. Infl racemose/fasciculate/fls solit. Fls uni/bisex, act. K hypog CA perig. K5/(5), C(5), A5 antipet extrorse poricidal, + 5 staminodes, G1-loc; ov n, free-central. Berry/drupe. *New World Trop.*

**99. Myrsinaceae.** Woody. Lvs alt, simple, exstip, with pellucid glands. Infl cymose/fasciculate. Fls uni/bisex, act. K hypog CA perig. K4–6/(4–6), C usu (4–6), A4–6 antipet introrse, G(4–6); ov n usu free-central. Berry/drupe. *Mainly Trop.*

**100. Primulaceae.***† Herbs rar shrublets/aquatic. Lvs alt/opp/basal, usu simple, exstip. Infl not cymose. Fls bisex, usu act. K hypog CA perig. K(4–9) usu (5), C(4–9) usu (5) rar 0, A5–9 antipet, G usu (5), stigma capitate; ov usu n, free-central. Capsule. *Widespread.*

## Superorder Rosidae

Flowers mostly polypetalous or apetalous, sometimes zygomorphic; stamens, when numerous, developing centrifugally; pollen binucleate or trinucleate; ovary sometimes apocarpous; placentation seldom parietal.

ROSALES

Woody/herbaceous; lvs often compound; carpels free/united.

**101. Eucryphiaceae.** Woody. Lvs opp, simple/pinnate, stip. Fls solit, bisex, act. KCA hypog. K4, C4, A n, G(5–12) rar (–18); ov n, ax. Capsule. *Chile, Australia, Tasmania* (*Eucryphia*).
**102. Cunoniaceae.** Woody. Lvs opp, usu pinnately compound, stip. Infl var. Fls usu bisex, act. KCA perig/epig. K3–6/(3–6), C3–5 rar 0, A n, G(2–5); ov n, ax. Capsule/nut. *Mostly S Hemisphere.*
**103. Pittosporaceae.** Woody. Lvs alt/opp, simple, exstip. Infl var. Fls bisex, usu act, KCA hypog. K5, C5, A5, G(2–5); ov n, ax. Capsule/berry. *Trop & S Temp Old World.*
**104. Droseraceae.***† Herbs. Lvs rosulate, usu simple, insectivorous with glandular hairs. Infl raceme/panicle. Fls bisex, act. KCA hypog. K(4–5), C5, A5–20, G(3–5); ov n, par. Capsule. Pollen in tetrads. *Temp.* (*Often placed in Sarraceniales.*)
**105. Crassulaceae.***† Usu leaf succulents. Lvs opp/alt, simple, exstip. Infl cymose. Fls bisex, act. KCA hypog. K3–30/(4–5), C3–30/(4–5), A3–n, G3–n or united below; ov n, marginal. Follicles. *Widespread* (*except Australia*).
**106. Saxifragaceae** s.1.*† Herbs/woody. Lvs usu alt, simple/compound, usu exstip. Fls usu bisex & act. KCA perig/epig rar hypog. K4–5/(4–5), C4–5 rar 0, A usu 3–10 rar –30, G(2) rar (–6), often with divergent styles; ov n, ax/par. Capsule/berry. *Widespread.*
**107. Rosaceae.***† Herbs/woody. Lvs usu alt & stip, simple/compound, often serrate. Infl var. Fls usu bisex & act. KCA perig/epig. K4–5 rar –9, free or united, sometimes with epicalyx, C0–5 rar –9, A n–4 rar fewer, G1–n/(2–5); ov 1–n, ax/marginal/basal. Follicles/achenes/drupe/pome. *Widespread.*

FABALES

Herbaceous/woody; roots often with nodules containing nitrogen-fixing bacteria; fr usu legume.
**108. Leguminosae/Fabaceae** s.1.*† Herbs/woody. Lvs usu alt, stip & pinnately compound, sometimes 2–3-foliolate/simple. Infl var. Fls usu bisex & zyg. KCA hypog/perig. K usu 5/(5), C5 free or variously connate, A n–4 free or variously connate, G1, rar –15; ov 1–n, marginal. Legume (sometimes indehiscent)/lomentum/nut. *Widespread.*

PODOSTEMALES

Much modified aquatic herbs.
**109. Podostemaceae.**† Aquatics of running water resembling algae, mosses or hepatics. Lvs alt, simple. Fls bisex, zyg. PA hypog. P2–3/(2–3), A1–4, G(2); ov n, ax. Capsule. *Mainly Trop.*

Mostly aquatic or marsh herbs; fls unisex, epig.

**110. Haloragaceae.***† Herbs. Lvs opp, simple/divided, exstip. Fls usu unisex, act. KCA/PA epig. K2–4, C0–4, A2–8, G(4); ov 1 per cell, ax. Nut/drupe. *Widespread.*

**111. Gunneraceae.** Herbs of marshy places. Lvs alt/basal, simple, stip, petiolate, sometimes very large. Infl racemose. Fls usu unisex, act. PA epig. P2–4, A2, G(2), 1–loc; ov 1, apical. Fr drupe/nut. *S Hemisphere, mainly S America* (*Gunnera*).

**112. Hippuridaceae.***† Aquatic. Lvs whorled, entire, exstip. Fls axillary, bisex, A epig. P0, A1, G1-loc; ov 1, apical. Cypsela. *Temp* (*Hippuris*).

**113. Theligonaceae.*** Fleshy herb. Lower lvs opp, upper alt, all simple with sheathing bases. Infl cymose. Male fls act, PA hypog, P2, A7–22; female fls ± zyg, P ± tubular, G1, style at last lateral; ov 1, basal. Nut. *Mediterranean, China, Japan* (*Theligonum – the family is now usually included in the Rubiales*).

MYRTALES

A n–few; G usu inf; ov ax/apical.

**114. Lythraceae.***† Usu herbs. Lvs opp, simple, stip/exstip. Fls bisex, usu act. KCA perig. K4–6–8 (epicalyx frequent), C4–8 rar 0, A6–16, rar fewer or more, often with unequal filaments, G(2–6); ov n, ax. Capsule. *Widespread.*

**115. Thymelaeaceae.***† Usu woody. Lvs alt/opp, simple, exstip. Infl var. Fls bisex, usu act. PA perig. K(4) rar (5), often petaloid, C0 rar 4, A2–8, G1; ov 1–2, ± apical. Drupe/nut/capsule. *Widespread.*

**116. Trapaceae.*** Aquatic herbs. Lvs opp, simple, exstip, with inflated petioles. Fls bisex, act. KCA ± epig. K4, C4, A4, G(2); ov 1 per cell. Horned drupe. *Old World* (*Trapa*).

**117. Myrtaceae.***† Woody. Lvs usu opp, simple, exstip, with pellucid, aromatic glands. Infl var. Fls bisex, act. KCA usu epig. K4–5/(4–5), C4–5/(4–5), A n/(n), G(3–n); ov 2–n, ax/par. Capsule/berry. *Mostly Trop America & Australia.*

**118. Punicaceae.** Woody. Lvs opp, simple, exstip. Infl cymose/fls solit. Fls bisex, act. KCA epig. K5–8, C5–7, A n, G usu (8–12) rar (3); ov n, ax. Fr berry-like. *Warm N Temp Old World, Socotra* (*Punica*).

**119. Onagraceae.***† Usu herbs. Lvs alt/opp, simple, stip/exstip. Fls solit/racemose, bisex, usu act. KCA usu epig. K(2–6) usu (4), C4 rar 2, A4–8 rar 1, G(1–5) usu (4); ov n, ax. Capsule/berry/nut. *Mostly Temp.*

**120. Melastomataceae.**† Woody/herbs. Lvs opp, exstip, simple. Infl cymose. Fls bisex, ± act. KCA perig/epig. K usu 4–5/(4–5), C4–5, A10–4, geniculate, poricidal, G usu (4–5); ov n, ax. Capsule/berry. *Mainly Trop.*

**121. Combretaceae.** Woody. Lvs opp, entire, exstip. Infl racemose. Fls usu bisex, act. KCA epig. K(4–5) rar (8), valvate, C4–5 rar –8/0, A4–10 rar n, G1-loc; ov 2–6, apical. Fr leathery, 1-seeded, often winged. *Trop.*

CORNALES

Usu woody; lvs simple, usu exstip; G inf; drupe/berry.

**122. Davidiaceae.** Trees. Lvs alt, simple, exstip. Infl a capitulum subtended by 2 showy white bracts, consisting of 1 bisex fl surrounded by many males. P0, A1–7, epig in bisex fl, G6–10-loc; ov 1 per cell, ax. Drupe. *China* (*Davidia*).

**123. Nyssaceae.**† Trees. Lvs alt, simple, exstip. Fls in axillary clusters, unisex, act. KCA epig. K(5), C5, A5–12, G1–2-loc; ov 1, ax. Drupe. *Temp N America, China.*

**124. Alangiaceae.** Trees/shrubs, sometimes spiny. Lvs alt, simple/lobed, exstip. Infl cyme. Fls bisex, act. KCA epig. K(4–10), C4–10, recurving, A4–n, G(2–3), 1-loc; ov 1, pendulous. Drupe. *Old World Trop & Subtrop* (*Alangium*).

**125. Cornaceae.***† Usu woody. Lvs alt/opp, simple, usu exstip. Infl cymes/panicles/capitula, rar racemes. Fls uni/bisex, act. KCA epig. K4–5/(4–5), C4–5 rar 0, A4–5, G(2–4); ov 1 per cell, ax/rar par. Drupe/berry. *Mostly Temp.*

**126. Garryaceae.**† Evergreen shrubs. Lvs opp, entire, exstip. Infl catkin. Fls unisex. Male: P4, A4; female: P0–4, G(2) naked/inf; ov 2, apical. Berry. *N America* (*Garrya*).

PROTEALES

Woody; apetalous; P usu 4; G1.

**127. Elaeagnaceae.***† Woody. Lvs alt/opp, simple, exstip, often with lepidote scales. Infl. var. Fls uni/bisex, act. PA perig. P(2–6) usu (4), A4–12 often alternating with P lobes, G1; ov 1, basal. Achene in fleshy P. *Widespread.*

**128. Proteaceae.** Usu woody. Lvs alt, exstip. Infl var. Fls usu bisex, act/zyg. PA perig. P(4), A4 borne on petaloid, spoon-shaped P segments, G1; ov 1–n, marginal. Follicle/nut/drupe. *S Hemisphere.*

SANTALALES

Often parasitic, usually green; C absent; G inf.

**129. Santalaceae.***† Habit var, some hemiparasitic. Lvs opp/alt, simple, exstip. Fls uni/bisex, act. PA epig. P4–5/(4–5), A4–5, G(3–5); ov 1–5, basal. Nut/drupe. *Widespread.*

**130. Loranthaceae.***† Mostly branch parasites. Lvs usu opp, simple, exstip, leathery. Fls uni/bisex, act. PA epig. P4–6/(4–6), A4–6, G(3–6); ov not differentiated in flower. Fr a 2–3-seeded berry/drupe. *Mostly Trop.*

**131. Cynomoriaceae.*** Root parasites without chlorophyll. Lvs scale-like. Infl spicate to capitate. Fls usu unisex, act. PA epig. P1–5, A1, G1; ov 1 ± apical. Fr small nut. *Mediterranean region, C Asia* (*Cynomorium*).

### RAFFLESIALES

Parasites, vegetatively very reduced; G inf; ov par.

**132. Rafflesiaceae.***† Root or branch parasites lacking chlorophyll. Lvs scale-like/0. Fls unisex, act. PA epig. P4–10/(4–10), A n, G (4–6–8); ov n, par. Berry. *Mostly Old World Trop.* (*Often placed in Aristolochiales.*)

### CELASTRALES

Woody; usu polypetalous; hypog disc often well-developed.

**133. Celastraceae.***† Woody. Lvs alt/opp, simple, stip/exstip. Infl cymes. Fls usu bisex, act, KCA hypog/perig, with disc. K(4–5), C4–5 rar –10, A4–5 rar –10, G(2–5); ov usu 2 per cell, ax. Fr var, seeds arillate. *Widespread.*

**134. Salvadoraceae.** Woody. Lvs opp, simple, with minute stip. Infl fasciculate/panicle. Fls uni/bisex, act. KCA hypog, disc 0. K(3–4), P4/(4) A4/(4), G1–2-loc; ov 1–2, basal. Berry/drupe. *Trop Africa & Asia.*

**135. Aquifoliaceae.***† Woody. Lvs alt, simple, exstip. Infl fasciculate/cymes. Fls uni/bisex, act. KCA hypog, disc 0; K(3–6), C4–5/(4–5), A4–5, G(3–n); ov 1–2 per cell, ax. Drupe. *Widespread.*

### EUPHORBIALES

Unisex; usu apetalous; G(2–4), usu 3-locular.

**136. Buxaceae.***† Evergreen, usu woody. Lvs usu opp, simple, exstip. Infl spike/raceme/fasciculate. Fls unisex, act. PA hypog. P(4–12) rar 0, A4–n, G(2–4) usu (3), styles undivided; ov usu 2 per cell, ax. Loculicidal capsule/berry-like, seeds shiny black. *Mostly Trop.*

**137. Euphorbiaceae.***† Woody/herbs/succulents, milky sap often present. Lvs usu alt & stip, simple/compound. Infl var. Fls unisex, act (sometimes borne in gland-bearing cup – cyathium). PA hypog. P4–5, 10, rar K5/(5), C5/(5), An–1, G(2–4) usu (3), styles often divided; ov

1–2 per cell, ax. Fr usu schizocarpic, seeds often carunculate. *Widespread.*

**138. Daphniphyllaceae.** Trees/shrubs. Lvs alt, crowded, entire, exstip, usu evergreen. Infl axillary raceme. Fls unisex, act. PA hypog. Male: P3–8, imbricate, A6–12; female: P0, staminodes few, small/0, G(2), imperfectly 2-loc, styles 1–2, undivided, persistent; ov 2 per cell, pendulous. 1-seeded drupe. *Temp E Asia* (*Daphniphyllum*).

RHAMNALES

Woody; A antipet; KCA often perig, disc usu present; ov 1–2 per cell.

**139. Rhamnaceae.***† Woody. Lvs usu alt & stip, simple. Infl corymb/cyme/fasciculate. Fls uni/bisex, act. KCA perig/epig, disc usu present. K5–4, C5–4, A5–4 antipet, G(2–4); ov 1, rar 2 per cell, ax. Capsule/drupaceous. *Trop, N Temp.*

**140. Vitaceae.***† Usu tendrillar climbers. Lvs alt, simple/compound, stip/exstip. Infl often cymose, leaf-opposed. Fls uni/bisex, act. KCA perig, with disc. K4–5/(4–5), C4–5/(4–5) often calyptrate, A4–5 antipet, G(2–6) usu (3); ov 1–2 per cell, ax. Berry. *Trop & warm Temp.*

SAPINDALES

Woody; A rar antipet; disc often present between perianth and ovary.

**141. Staphyleaceae.***† Woody. Lvs usu opp, compound, stip. Infl raceme/panicle. Fls usu bisex, act. KCA perig with disc. K(5), C5, A5, G(2–3); ov n, ax. Inflated capsule. *Trop & N Temp* (*disjunct*).

**142. Melianthaceae.** Usu woody. Lvs alt, pinnate, stip intrapetiolar. Infl raceme. Fls bisex, zyg. KC perig, A hypog, with disc. K5 rar 4, C4–5, A4–5 rar 10, free/united, G(4–5); ov 1–n per cell, ax. Capsule. *Africa.*

**143. Greyiaceae.** Woody. Lvs alt, lobed, exstip. Infl racemose. Fls bisex, act. KCA hypog with cupular 10-lobed disc. K5, C5, A10, G(5), style 1; ov n, par. Capsule. *S Africa* (*Greyia*).

**144. Sabiaceae.** Woody. Lvs alt, simple/compound, exstip. Infl panicle. Fls usu bisex, zyg, KCA hypog, disc small. K3–5/(3–5), C4–5, A3–5 antipet, G(2); ov 2 per cell, ax. Berry. *Mostly Trop.*

**145. Sapindaceae.**† Usu woody. Lvs usu alt & compound & exstip. Infl var. Fls uni/bisex, act/zyg. KCA usu hypog, disc outside A. K5, C4–5 rar 0, A4–n often 8, G(3) rar 1–(4); ov 1–2 per cell, ax. Fr var, seeds often arillate. *Mainly Trop.*

**146. Hippocastanaceae.***† Woody. Lvs opp, palmate, exstip. Infl racemose. Fls usu bisex, zyg. KC perig A usu hypog, disc present. K(4–5), C4–5, A5–9, G(3); ov 2 per cell, ax. Capsule, large-seeded. *Trop & Temp.*

**147. Aceraceae.***† Woody. Lvs usu opp, simple/compound, exstip. Fls clustered, uni/bisex, act. KC perig A hypog/perig, disc present. K4–5, C4–5 rar 0 A4–5, G(2–3); ov 2 per cell, ax. Winged mericarps. *N Temp.*

**148. Burseraceae.**† Woody; aromatic resins. Lvs alt, usu compound, exstip. Infl panicles/fls solit. Fls uni/bisex, act. KCA hypog with disc. K(3–5), C3–5 rar 0, A6–10, G(2–5); ov 2 per cell, ax. Drupe/capsule. *Mainly Trop.*

**149. Anacardiaceae.***† Woody, resinous. Lvs usu alt, simple/compound, exstip. Infl panicle. Fls usu bisex & act. KCA hypog, disc often present. K(3–5),C3–5 rar 0, A3–10, G usu (3) rar (1–5); ov 1 per cell, apical/basal. Drupe, 1-seeded. *Trop & warm Temp.*

**150. Simaroubaceae.**† Woody. Lvs alt, simple/compound, exstip. Infl var. Fls usu unisex, act. KCA hypog, disc present. K(3–8), C0–8, A6–14 rar more, G(2–5); ov 1–2 per cell, ax. Fr var. *Mainly Trop.*

**151. Cneoraceae.*** Shrubs, sometimes with medifixed hairs. Lvs alt, simple, exstip. Infl cyme. Fls bisex, act. KCA hypog, disc 0. K3–4, C3–4, A3–4, G(3–4) on gynophore; ov 1–2 per cell, ax. Schizocarp. *Mediterranean, Canary Is, Cuba.*

**152. Coriariaceae.*** Usu shrubs, branches angular. Lvs opp, entire, exstip. Infl raceme. Fls usu bisex, act. KCA hypog. K5, C5 keeled inside, A10, G5–10; ov 1, apical. Achenes surrounded by fleshy C. *Scattered* (*Coriaria – position doubtful*).

**153. Rutaceae.***† Woody/herb. Lvs alt/opp, simple/compound, exstip, usu aromatic, gland-dotted. Infl var. Fls usu bisex, act. KCA usu hypog, disc usu present. K3–5/(3–5), C3–5/(3–5) rar 0, A3–10, G4–5 rar n, united at least by a common style, often on short gynophore; ov 1–n, ax. Fr fleshy/capsular/samara. *Trop, warm Temp.*

**154. Meliaceae.** Trees, wood often scented. Lvs usu alt, mostly pinnate, exstip. Infl cymose panicle. Fls usu bisex, act. KCA hypog, disc present. K(4–5), C4–5 rar –8, A(8–10) rar free or fewer/more numerous, G(2–5); ov usu 2 or more per cell, ax. Berry/capsule/drupe. *Mainly Trop.*

**155. Zygophyllaceae.***† Herbs/shrubs. Lvs usu opp, usu compound, stip, often fleshy. Infl cymes/fls solit. Fls usu bisex & act. KCA hypog, disc usu present. K4–5, C4–5, A5–15, G(2–5); ov n, ax. Capsule/drupaceous/schizocarp. *Trop & warm Temp.*

## JUGLANDALES

Lvs pinnately compound; infl catkin; monoecious.

**156. Juglandaceae.***† Woody. Lvs usu opp, pinnately compound, exstip. At least male infl catkin. Fls unisex, P4, A3–n, G(2–3), inf; ov 1, basal. Nut with complex lobed and folded cotyledons. *N Temp.*

### GERANIALES

Mostly herbs; lvs simple/divided; G superior; dehiscence of fruit often violent.

**157. Linaceae.***† Herbs/shrubs. Lvs alt/opp, entire, stip/exstip. Infl cyme. Fls bisex, act. KCA hypog. K5–4/(5–4), C5–3, A5–4 rarely –20, G(5–3) often 10–6-loc with 5–3 secondary septa; ov 1–2 per cell, ax. Capsule/drupe. *Widespread.*

**158. Erythroxylaceae.** Woody. Lvs alt, simple, stip. Infl var. Fls bisex, act. KCA hypog. K5, C5 appendaged on inner face, A(10), G(3), often only 1 cell developing; ov 1–2 per cell, ax. Fr berry-like. *Trop (mostly America).*

**159. Geraniaceae.***† Usu herbs. Lvs alt/opp, simple/compound, stip/exstip. Infl var. Fls bisex, act/zyg. KCA usu hypog. K3–5/(3–5), C3–5/(5), A5–15/(5–10), G(3–5), sup, often long-beaked; ov 1–n per cell, ax. Capsule/berry/schizocarp. *Widespread.*

**160. Oxalidaceae.***† Mostly herbs. Lvs alt/basal, trifoliolate/digitate/pinnate, exstip. Infl var. Fls bisex, act. KCA hypog. K5/(5), C5 contorted, A10, G(5) with free styles; ov 1–more, ax. Capsule, often explosive. *Widespread.*

**161. Limnanthaceae.**† Herbs. Lvs alt, divided, exstip. Fls solit, bisex, act. KCA hypog. K3–5, C3–5, A6–10, G3–5 united by a gynobasic style; ov 1 per cell, ascending. Nutlets. *Temp N America.*

**162. Tropaeolaceae.** Herbs. Lvs alt/opp, simple/divided, exstip. Fls bisex, zyg, solit, axillary. KC partly perig, A hypog. K5, C5, A8, G3-lobed, style 1; ov 1 per cell, ax. Schizocarp. *C & S America (Tropaeolum).*

**163. Balsaminaceae.***† Herbs. Lvs alt/opp, simple, exstip. Infl var. Fls bisex, zyg. KCA hypog. K3 rar 5, often coloured, the lowest spurred; C5 upper exterior, laterals united, A(5), G(5); ov n, ax. Explosive capsule. *Mostly Old World.*

### POLYGALALES

Herbs/shrubs; fls mostly zyg; A often poricidal.

**164. Malpighiaceae.**† Woody, often with medifixed hairs. Lvs usu opp, simple, stip. Infl var. Fls usu bisex, act. KCA hypog. K5 often with 2 glands on back, C5 petals often fringed, A(10), G(3); ov 1 per cell, ax. Fr var, often winged mericarps. *Mainly Trop America.*

**165. Tremandraceae.** Shrublets. Lvs usu opp, entire, exstip. Fls solit, bisex, act. KCA hypog. K4–5, C4–5, A8–10 poricidal, G(2); ov 1–2 rar 3 per cell, ax. Capsule. *Australia.*

**166. Polygalaceae.***† Herbs/shrubs. Lvs usu alt, entire, exstip. Infl racemose. Fls bisex, zyg, papilionaceous. KCA hypog/K hypog CA perig. K usu 5, lateral pair petaloid, C usu 3, often adnate to staminal tube, A(10–8), poricidal, G usu (2); ov usu 1 per cell, ax. Usu capsule, seeds arillate. *Widespread.*

**167. Krameriaceae.**† Shrubs/herbs. Lvs alt, entire, exstip. Infl axillary/ racemose. Fls bisex, zyg, papilionaceous. KCA hypog/K hypog CA perig. K4–5, C5 free or united, heteromorphic, the abaxial pair often modified into glands, A3–4 poricidal, G1-loc; ov 2, pendulous. Fr 1-seeded, indehiscent, covered with barbed spines. *Mainly Trop America* (*Krameria*).

UMBELLALES

Fls polypetalous, usu umbellate; G inf; ov ax.

**168. Araliaceae.***† Usu woody. Lvs alt, usu lobed/compound, stip, stellate hairs frequent. Infl usu umbellate. Fls usu unisex, act. KCA epig. K5/(5), C5–10 valvate, A5–10, G(2–15); ov 1 per cell, ax. Berry/drupe. *Mostly Trop.*

**169. Umbelliferae/Apiaceae.***† Usu herbs. Lvs alt, often pinnately compound, petioles sheathing. Infl umbel, rar head. Fls usu bisex, act/zyg. KCA epig. K(5) often very reduced, C5 imbricate, inflexed, A5, G(2); ov 1 per cell, ax. Schizocarp. *Mainly N Hemisphere.*

## Superorder Asteridae

Flowers actinomorphic or zygomorphic, gamopetalous; stamens never numerous; ovary always syncarpous; pollen binucleate or trinucleate.

GENTIANALES

Lvs entire or sometimes pinnate; fls act; A4–5, G(2) sup.

**170. Loganiaceae.**† Woody, sometimes climbers, with internal phloem; glandular hairs absent. Lvs opp, entire/spinose, stip. Infl cymose/fls solit. Fls bisex, act. K hypog CA perig. K(4–5) imbricate, C(4–5) imbricate/contorted, A4–5, G(2); ov n, ax. Capsule/berry/drupe. *Trop.*

**171. Gentianaceae.***† Herbs. Lvs opp, entire, exstip. Infl cymose/fls solit. Fls bisex, act. K hypog CA perig. K4–5 rar 12, usu united, C(4–5) rar (–12), contorted, A4–5 rar –12, G(2); ov n, usu par. Capsule. *Mainly Temp & Subtrop.*

**172. Apocynaceae.***† Woody/herbs, often climbing, with milky sap. Lvs entire, usu opp, exstip. Infl racemose/cymose/fls solit. Fls bisex, act. K hypog CA perig. K5–4/(5–4), C(5) contorted, A5, G(2) often united only by style; ov n, marginal/ax. Fr var, seeds often plumed. *Widespread, centred in Trop.*

**173. Asclepiadaceae.***† Woody/herbs/climbers, usu with milky sap. Lvs opp, entire, stip minute/0. Infl racemose/cymose/fls solit. Fls bisex, act. K hypog CA perig. K5/(5), C(5) contorted/valvate, corona frequent, A5 often adnate to style, 'translators' and pollinia frequent, G(2), often united only by style; ov n, marginal. Fr 1–2 follicles, seeds plumed. *Mostly Trop.*

**174. Oleaceae.***† Woody, sometimes climbing. Lvs usu opp, simple/pinnately compound, exstip. Infl often cymose panicle. Fls bisex, act. KCA hypog/K hypog CA perig. K(4) rar 0–(15), C(4) rar 0–(15), A2, anther cells back to back, G(2); ov usu 2 per cell, ax. Fr var. *Temp & Trop.*

POLEMONIALES

Fls mostly act; A5 rar fewer; G(2–3) rar more, sup.

**175. Nolanaceae.** Herbs/shrublets. Lvs alt, simple, exstip, fleshy. Fls axillary, bisex, act. K hypog CA perig. K(5), C(5) infolded in bud, A5, G(5), lobed; ov few, ax. Schizocarp. *Chile, Peru.*

**176. Solanaceae.***† Woody/herbs, with internal phloem. Lvs alt, simple rar pinnatisect, exstip. Infl often cymose/fls solit, often extra-axillary. Fls bisex, act/zyg. K hypog CA perig. K5/(5), C(5) lobes folded/contorted/ valvate, A5 rar 4/2, G usu (2), septum usu oblique, rar with secondary septa; ov n, ax. Berry/capsule. *Widespread.*

**177. Convolvulaceae.***† Climbers/shrublets, often with milky sap. Lvs alt, simple, exstip. Infl cymose/fls solit. Fls bisex, act. K hypog CA perig. K5, C(5) contorted and usu trumpet-shaped, rar imbricate-lobed, A5, G(2), 2–4-loc; ov 1–2 per cell. Capsule/fleshy. *Widespread.*

**178. Cuscutaceae.***† Twining stem parasites, chlorophyll 0. Lvs scale-like. Fls clustered, bisex, ax. K hypog CA perig. K4–5/(4–5), C(4–5), A5 with scales below them, G(2); ov 4 ax/par. Capsule. *Widespread (Cuscuta).*

**179. Menyanthaceae.***† Aquatic/marsh herbs. Lvs alt, entire/tri-foliolate, petioles sheathing. Infl var. Fls bisex, act. K hypog CA perig. K5/(5), C(5) valvate, A5, G(2); ov n, par. Usu capsule. *Temp.* (*Often placed in Gentianales*.)

**180. Polemoniaceae.***† Herbs rar shrubs/tendrillar climber. Lvs alt/opp, entire/pinnately divided, usu exstip. Infl cymose to capitate rar fls solit. Fls bisex, usu act. K hypog CA perig. K(5), C(5) contorted, A5, G(3); ov 1–n per cell, ax. Capsule. *Mainly America.*

**181. Hydrophyllaceae.**† Usu herbs. Lvs alt/basal rar opp, entire/ divided, exstip. Infl scorpioid cymes/fls solit. Fls bisex, act. K hypog CA perig. K5/(5), C(5) usu imbricate, A5, G(2); ov n rar –4, ax/par. Capsule. *Mainly America.*

**182. Boraginaceae.***† Herbs/woody. Lvs alt, simple, exstip. Infl often scorpioid cyme. Fls bisex, act rar zyg. K hypog CA perig. K5/(5), C(5), A5, G(2) usu 4-celled by secondary septation; ov 4 side by side; style terminal/gynobasic. Fr 4 (rar 1) nutlets/drupe. *Widespread.*

**183. Lennoaceae.**† Parasitic herbs, chlorophyll 0. Lvs scale-like. Infl spicate/cymose/capitate. Fls bisex, act. K hypog CA perig. K(6–10), C(5–8) imbricate, A5–8, G(6–15); ov 2 per cell, ax. Fleshy capsule. *SW USA, Mexico.*

LAMIALES

Stem often 4-angled; lvs opp; fls usu zyg; A usu 4/2; G(2) sup; fr fleshy/2–4 nutlets.

**184. Verbenaceae.***† Woody/herbs. Lvs opp, simple/compound, exstip. Infl var. Fls bisex, zyg. K hypog CA perig. K(5–8) ± act, C(5), A4 rar 2–5, G2–9-loc, style terminal; ov 1–2 per cell, ax rar par. Drupe/berry/ rar 4 nutlets. *Mainly Trop.*

**185. Labiatae/Lamiaceae.***† Herbs/shrubs. Lvs opp, aromatic. Infl often verticillate. Fls bisex, zyg. K hypog CA perig. K usu (5) often zyg, C(5) rar (3), 1–2-lipped, A4/2, G(2) 4-celled by secondary septation; style gynobasic/terminal; ov 1 per cell, ax. Fr 4 nutlets rar fleshy. *Widespread, centred in Mediterranean region.*

**186. Phrymaceae.**† Herbs. Lvs opp. Infl spicate. Fls bisex, deflexed in fr, zyg. K hypog CA perig. K(5) teeth hooked, C(5), A4, G(2); ov 1, basal. Nut in persistent K. *E Asia, Atlantic N America* (*Phryma*).

**187. Callitrichaceae***† Aquatic herbs. Lvs opp, simple, exstip. Fls solit, axillary, unisex, act, A hypog. P0, A1, G(2) 4-celled by secondary septation; ov 1 per cell, ax. Schizocarp. *Widespread* (*Callitriche*).

PLANTAGINALES

Fls act, 4-merous; C scarious, small; A exserted; G sup.

**188. Plantaginaceae.***† Herbs rar shrublets. Lvs alt/opp/basal. Infl usu spicate. Fls uni/bisex, act. K hypog CA perig/KCA hypog. K4/(4), C(4–3), A usu 4, G1–4-loc; ov few. Capsule opening by lid/nut. *Widespread.*

SCROPHULARIALES

C often 2-lipped; A4/2; G(2) sup; fr usu capsule/rar 1-seeded.

**189. Buddlejaceae.** Woody rar herbaceous, without internal phloem, often with glandular hairs. Lvs opp/whorled, rar alt, often toothed, stip forming a line uniting bases. Infl var. Fls bisex, act. K hypog CA perig. K(4), C(4), A4, G(2), style 1; ov n, ax. Capsule/berry/drupe. *Mainly Trop E Asia.* (*Often placed with Loganiaceae.*)

**190. Scrophulariaceae.***† Herbs/woody some hemiparasitic, internal phloem absent. Lvs alt/opp, simple rar compound. Infl var. Fls bisex, usu zyg. K hypog CA perig. K(4–5), C(4–5) rar (–8) lobes imbricate, A4/2 rar 5, G(2), septum horizontal; ov n–1 ax, placentae usu simple. Capsule/rar berry or indehiscent. *Widespread.*

**191. Myoporaceae.** Woody. Lvs usu alt, often with resinous glands. Infl var. Fls bisex, usu zyg. K hypog CA perig. K(5), C(5), A4 rar 5, G(2); ov 4–8, ax. Fr drupe-like. *Scattered, chiefly Australasian.*

**192. Globulariaceae.*** Herbs/shrublets. Lvs alt/rosulate. Infl bracteate capitulum. Fls bisex, zyg. K hypog CA perig. K(5), C(4–5), A4, G(2), 1-loc; ov 1, apical. Nut. *Mostly Mediterranean.*

**193. Gesneriaceae.*** Herbs/shrubs, some epiphytic. Lvs usu opp/basal, often velvety. Infl cymose/fls solit. Fls bisex, usu zyg. K hypog CA perig/KCA epig. K(5), C(5), A4/2 rar 5, G(2); ov n, par, placentae intrusive, bifid. Capsule/berry. *Mostly Trop.*

**194. Orobanchaceae.***† Parasitic herbs, chlorophyll 0. Lvs alt, scale-like. Fls bisex, zyg. K hypog CA perig. K(4–5), C(5), A4, G(2) rar (3); ov n, par, usu on 4 placentae. Capsule. *Mainly N Temp* (*Lathraea, often placed in Scrophulariaceae, is included here.*)

**195. Bignoniaceae.**† Usu woody/climbing. Lvs usu opp, compound rar simple. Infl usu cymose. Fls bisex, zyg. K hypog CA perig. K(5), C(5), A4 rar 2, G(2); ov n, ax rar par. Usu capsule, seeds winged. *Mainly Trop, centred in S America.*

**196. Acanthaceae.***† Usu herbs. Lvs opp, simple, often with cystoliths. Infl cymose, often conspicuously bracteate. Fls bisex, zyg. K hypog CA perig. K(4–5), C(5) bilabiate, A4/2, G(2); ov ax, 2 or more per cell, often superposed. Fr usu explosive capsule. *Mainly Trop.*

**197. Pedaliaceae.**† Herbs, often viscid-pubescent. Lvs opp/alt above, simple, exstip. Infl racemes/axillary cymes/fls solit. Fls bisex, zyg. K hypog CA perig. K(5), C(5), A4 rar 2, G(2) 2–4-loc; ov 1–n per cell. Capsule, often 2-horned/nut-like. *Trop, S Africa.*

**198. Lentibulariaceae.***† Herbs, mostly insectivorous, some aquatic. Lvs alt/basal, often dimorphic, elaborated. Infl scapose, raceme/fls solit. Fls bisex, zyg. K hypog CA ± perig. K2–5/(2–5), C(5) spurred at base, A2, G(2); ov n, free-central. Capsule. *Widespread.*

Lvs alt; A5 rar –2, often free from C & convergent; G usu inf; ov usu n.

**199. Campanulaceae.***† Herbs, often with milky sap, rar woody. Lvs usu alt, simple, exstip. Infl var. Fls bisex, act/zyg. KCA epig/rar hypog. K5 rar 3–10, C(5) rar (3–10), valvate, rar ± free, A5 rar 3–10, rar shortly epipet, G(2–5) rar (–10); ov n, ax. Capsule/fleshy. *Widespread.*

**200. Stylidiaceae.** Herbs. Lvs basal/cauline, linear, usu exstip. Infl var. Fls uni/bisex, act/zyg. KCA epig. K(5–7), C(5) imbricate, A2 adnate to style, G(2); ov n, ax/par/free-central. Fr usu capsule. *Australasia.*

**201. Brunoniaceae.** Herbs. Lvs basal, simple, exstip. Infl bracteate capitulum. Fls bisex, ± act. K hypog CA perig. K(5), C(5) valvate, A5 syngenesious, G1–loc; ov 1, basal. Nut enclosed in K tube. *Australia (Brunonia).*

**202. Goodeniaceae.** Herbs/shrubs. Lvs usu alt, simple, exstip. Fls bisex, zyg. KCA usu epig. K5/(5), C(5) 1–2-lipped, valvate/infolded, A5 free/epipet, G(2) 1–2-loc, stigma indusiate; ov 1–2 per cell, ax/basal. Fr var. *Mainly Australasia.*

RUBIALES

Mostly woody; lvs opp, stip usu present, often inter/intrapetiolar; G inf.

**203. Rubiaceae.***† Woody/herbs. Lvs opp, simple, stip (sometimes appearing whorled). Infl cymose/capitate/fls solit. Fls bisex, usu act, KCA usu epig. K4–5/(4–5), C(4–5) rar (–10), A4–5 rar 10, epipet, G(2 or more); ov 1–n per cell, ax. Capsule/berry/2 mericarps. *Widespread.*

DIPSACALES

Woody/herbaceous; lvs exstip; G inf.

**204. Caprifoliaceae.***† Mostly shrubs/climbers. Lvs opp, usu simple (rar pinnate) & exstip. Infl often cymose. Fls bisex, act/zyg, often twinned. KCA epig. K5/(5), C usu (5), A4–5 epipet, G(3–5) sometimes only 1 cell fertile; ov 1–n per cell, ax/pendulous. Berry. *Widespread, mainly N Temp.*

**205. Adoxaceae.***† Rhizomatous herb. Lvs opp/basal, compound, exstip. Fls in a head, bisex, act. KCA ± epig. K(2–3), C(4–6), A4–6 epipet, each split into 2 ½-anthered portions, G(3–5); ov 3–5, ax. Drupe. *N Temp (Adoxa).*

**206. Valerianaceae.***† Herbs. Lvs opp, simple/dissected, exstip. Infl cymose. Fls bisex, zyg. KCA epig. K tardily developing, C(5) often saccate/spurred, A1–4 epipet, G(3), 1 cell fertile; ov 1, pendulous. Cypsela, K often elaborated in fruit. *Mainly N Temp, Andes.*

**207. Dipsacaceae.*** Mostly herbs. Lvs opp, simple/dissected, exstip. Infl involucrate capitulum, rar fls in spiny-bracted whorls. Fls bisex, zyg. KCA epig with cupular involucel. K5–10 or cupular, C(4–5), A4 rar 2, epipet, G(2); ov 1, apical. Cypsela enclosed in involucel. *Old World, centred in Mediterranean.*

ASTERALES

Fls in involucrate capitula; stamens 5 usu syngenesious; G inf; ov 1, basal.

**208. Compositae/Asteraceae.***† Herbs/woody, sometimes with milky juice. Lvs var, exstip. Infl involucrate capitulum (rar 1-flowered). Fls uni/bisex, act/zyg. KCA epig. K reduced, C(5/3), A(5) rar 5, G(2); ov 1, basal. Cypsela, usu with pappus. *Widespread.*

## SUBCLASS MONOCOTYLEDONES

### Superorder Alismatidae

Flowers mostly apocarpous, some syncarpous with diffuse-parietal placentation; pollen grains trinucleate; endosperm usu absent, never starchy; stomata usually with 2 subsidiary cells; mostly aquatic.

ALISMATALES

Fls act; G sup, apocarpous; aquatics.

**209. Butomaceae.*** Aquatics without latex. Lvs basal, linear. Infl scapose bracteate umbel. Fls bisex, act. P3 + 3, petaloid, persistent, A9, G6 connate towards base, sup; ov n, diffuse-par. Fr ± follicular. *Temp Eurasia* (*Butomus*).

**210. Limnocharitaceae.** Aquatics with latex. Lvs with petiole & lamina. Fls solit/umbellate, bisex, act. K3, C3 not persistent, A6–n, G6–n, sup; ov diffuse-par. Follicles. *Trop.*

**211. Alismataceae.***† Scapose aquatics without latex. Lvs often broad. Infl usu much branched, rar 1 umbel. Fls uni/bisex, act. K3, C3, A6–n, G6–n, sup; ov 1–2, basal/marginal. Achenes. *Temp & Trop.*

HYDROCHARITALES

Fls act; P differentiated; G inf; ov diffuse-par; aquatic.

**212. Hydrocharitaceae.***† Aquatics with at least fls usu emergent. Lvs var. Fls (rar solit) arranged in a bifid spathe or between 2 opposite bracts, uni/bisex. K3, C3, A n–1, G usu (3–6), inf; ov n, diffuse-par. Capsule, rar berry-like. *Mainly Trop & warm Temp.*

P, when present, usu small & undifferentiated, rar 1–3 and petaloid; G sup/naked; often apocarpous; mostly aquatic (some marine), few marsh or bog plants.
**213. Aponogetonaceae.** Fresh water aquatics. Lvs long-petioled, sheathing. Infl a simple/forked spike. Fls usu bisex. P1–3/0, sometimes petaloid and (when P1) bract-like, A6 or more, G3–6, sup; ov few, basal. Follicles. *Old World, mainly Trop.*
**214. Scheuchzeriaceae.***† Bog herbs. Lvs distichous with ligulate sheaths. Fls in bracteate racemes, bisex, act. P6, A6, G3–6, sup; ov 2, basal. Follicles. *Cold N Temp* (*Scheuchzeria*).
**215. Juncaginaceae.***† Usu marsh plants. Lvs basal, sheathing. Infl ebracteate raceme/spike. Fls uni/bisex, act. P6 rar 1 (when sometimes interpreted as a bract), A1/6–4, G(3–6) sup, 1-loc; ov 1 per cell. Fr capsule/dimorphic & indehiscent. *Mainly Temp & cold regions; Pacific America.*
**216. Najadaceae.***† Submerged aquatics of fresh/brackish water. Lvs opp/whorled, entire/toothed. Fls solit at base of branches, unisex. Male: P2-lipped, A1; female: P membranous/0, G1-loc, sup/naked, with 2–4 stigmas; ov 1, basal. Fr indehiscent. *Widespread* (*Najas*).
**217. Potamogetonaceae.***† Submerged or emergent aquatics of fresh water. Lvs alt/opp, sheathing at base. Fls in axillary ebracteate spikes, bisex, act. P4, A4 inserted on P claw, G4 sup, each carpel with sessile stigma; ov 1, basal. Drupe/achene. *Widespread.*
**218. Ruppiaceae.***† Aquatics of saline marshes. Lvs opp/alt, linear, sheathing. Fls in 2-flowered spike, bisex, act. P0, A2, G4 or more, sup, stigmas dilated; ov 1, pendulous. Fr indehiscent, long-stalked, peduncle much elongating. *Temp & Subtrop* (*Ruppia*).
**219. Zannichelliaceae.***† Submerged aquatics of fresh/saline water. Lvs alt/opp/whorled, entire. Fls unisex in axillary cymes/solit. P cupular/3 scales/0, A3–1, G1–9, sup/naked, stigmas dilated/2–4-lobed; ov 1, pendulous. Fr indehiscent, stalked. *Widespread.*
**220. Zosteraceae.***† Submerged marine rhizomatous perennials. Lvs in 2 rows, sheathing at base & with stipuloid margins. Fls on flattened axis at first enclosed in lf sheath, unisex. P0 or represented by a row of bract-like lobes on each side of the axis. Male: A1; female: G1-loc, naked; ov 1, pendulous. Achene. *Widespread.*
**221. Posidoniaceae.*** Submerged marine perennials with fibrous rhizome. Lvs ligulate. Infl pedunculate, fls in spikes subtended by short lvs, bisex. P0/3 deciduous scales, A3–4, G1-loc, sup/naked, stigma sessile; ov 1. Drupe. *Mediterranean, Australasia* (*Posidonia*).

## Superorder Commelinidae

Flowers mostly syncarpous; perianth well differentiated into petals and sepals, or much reduced; pollen grains 2–3-nucleate; endosperm usually starchy; stomata with 2 or more subsidiary cells; plants rarely aquatic; leaves sometimes distichous.

### COMMELINALES

K & C strongly differentiated; staminodes present/0.

**222. Xyridaceae.**† Terrestrial/marsh plants with mostly radical lvs. Infl bracteate head. Fls bisex, ± zyg. K usu 3, segments heteromorphic, C(3), A3, staminodes 3–0, G(3), sup; ov few–n, par. Capsule. *Mainly Trop.*

**223. Mayacaceae.**† Aquatics. Lvs cauline, slender, apex 2-toothed. Fls axillary, bisex, act. K3, C3, A3 poricidal, G(3), sup; ov several, par. *Trop America, Africa* (*Mayaca*).

**224. Commelinaceae.**† Terrestrial herbs. Lvs mostly cauline, often with closed basal sheaths. Infl panicle/cincinnus/fls solit. Fls bisex, act/weakly zyg. K3, C3 rar fewer, A6–3 anthers basifixed, filaments sometimes hairy, staminodes 3–0, G(3), sup; ov few, ax. *Trop & warm Temp.*

### ERIOCAULALES

Fls minute, unisex, borne in capitula.

**225. Eriocaulaceae.***† Usu marsh plants. Lvs basal/cauline, sheathing. Fls in involucrate capitula, unisex, act/zyg. P usu biseriate, 4–6/(4–6) scarious/membranous, A4–6/2–3, G(2–3) sup; ov 1 per call. *Mainly Trop.*

### RESTIONALES

Lvs usu cauline, often reduced; fls in panicles/dioecious spikelets; P often scarious.

**226. Flagellariaceae.** Erect/climbing herbs. Lvs with closed sheaths, often distichous, sometimes ending in a tendril. Infl panicle. Fls uni/bisex, act. P6 dry, ± petaloid, A6, G(3), sup, style 3-lobed; ov 1 per cell, ax. Fr indehiscent. *Old World Trop.*

**227. Restionaceae.** Monoecious herbs. Lvs sheathing, usu reduced to sheaths. Fls in spikelets (1–n-flowered), unisex. P3/6 rar 0, scarious, A3/2, G1–3-loc, sup; ov 1 per cell. Capsule/nut-like. *Mainly S Hemisphere.*

### POALES

Fls distichously imbricated in spikelets; P reduced to 2/3 lodicules; fr a caryopsis.

**228. Gramineae/Poaceae.***† Herbs/bamboos. Lvs distichous (phyllotaxis $\frac{1}{2}$), sheathing and usu ligulate; stems round, internodes usu hollow. Fls compressed between bract (lemma) and bracteole (palea, rar 0), the unit comprising a floret, distichously arranged in spikelets subtended by usu 2 empty bracts (glumes). P represented by 2–3 lodicules, A3 rar 2/6, G1-loc, sup, styles 2 rar 3/1; ov 1, usu lateral. Seed adnate to pericarp (caryopsis). *Widespread.*

### JUNCALES

Fls clustered but not imbricated into regular spikelets; P6, scarious, G(3), ov 3–n; capsule.

**229. Juncaceae.***† Lvs often mostly basal, spirally arranged, sometimes reduced to sheaths. Stems round. Infl cymes/panicles/corymbs/heads. Fls usu bisex, act. P6 scarious, A usu 6, pollen in tetrads, G(3), sup; ov 3–n, ax/par. Capsule. *Widespread.*

### CYPERALES

Fls subtended by membranous bracts arranged spirally/distichously in spikes/spikelets; P reduced; fr nut-like.

**230. Cyperaceae.***† Herbs. Stems round/3-sided, usu solid. Lvs spirally arranged (phyllotaxis ± $\frac{1}{3}$), with closed sheaths. Fls subtended by membranous bracts (glumes), spirally/distichously arranged in spikes/spikelets without 2 empty glumes at base, bisex/unisex. P scales/bristles/hairs/0, A6 rar 3, with basifixed anthers, G1-loc, sup/naked, sometimes surrounded by a flask-shaped structure, styles 2–3; ov 1, basal. Fr nut-like. *Widespread.*

### TYPHALES

Aquatic/marsh herbs; lvs distichous, sheathing; fls unisex in dense spikes/heads; ov 1, apical.

**231. Sparganiaceae.***† Aquatic. Fls in unisex globose heads. P few scales, A3 or more, G1-loc, sup, ± sessile; ov 1, apical. Fr drupaceous. *N Temp, Australasia* (*Sparganium*).

**232. Typhaceae.***† Marsh plants. Infl of 2 unisex dense superimposed spikes. P threads/scales, A2–5, G1-loc, sup, on hairy stalk; ov 1, apical. Fr dry. *Widespread* (*Typha*).

### BROMELIALES

Lvs usu in basal rosettes, stiff & channelled; P strongly differentiated; septal nectaries frequent.

**233. Bromeliaceae.**† Herbs sometimes epiphytic. Lvs mainly basal, often spiny-margined or with elaborate trichomes. Infl usu terminal, conspicuously bracteate. Fls bisex, usu act. K3, C3/(3), A6 anthers usu versatile, G(3), sup/inf; ov n, ax. Berry/capsule. *Mainly Trop America.*

ZINGIBERALES

Lvs usu petiolate with broad blades pinnately parallel-veined; fls zyg/asymmetric, A6–1, G inf, septal nectaries frequent.

**234. Strelitziaceae.** Lvs distichous, sometimes huge. Infl erect cincinnus in axil of concave spathe. Fls bisex, zyg. K3 or adnate to C(3), the lower 2 C segments forming a sagittate structure, A5, G(3), inf; ov n, ax. Capsule. *Trop America, S Africa, Madagascar.*

**235. Lowiaceae.** Lvs radical, distichous. Fls in cymes arising from base of leaf sheaths, bisex, zyg. K3 united below into long tube, C3 lower petal forming a large labellum, A5, G(3), inf, style with 3 laciniate lobes; ov n, ax. Capsule. *Trop SE Asia* (*Orchidantha*).

**236. Heliconiaceae.** Lvs distichous. Infl a coloured terminal spathe bearing erect cincinni. Fls bisex, zyg. P3 + 3, petaloid, posterior segment large and free, remainder smaller and mostly connate, A5, G(3), inf; ov 1 per cell. Fr schizocarpic. *Trop America* (*Heliconia*).

**237. Musaceae.** Lvs huge, spirally arranged with inrolled petioles forming stem-like structure. Infl raceme. Fls subtended by large spathes, uni/bisex, zyg. K tubular, splitting down 1 side, C2-lipped, A5 + 1 staminode, G(3), inf; ov n, ax. Fr fleshy, often indehiscent, borne in trusses. *Old World Trop.*

**238. Zingiberaceae.** Lvs distichous, usu with open sheaths, ligulate, aromatic. Fls in racemes/heads/cymes, bisex, zyg. K(3), C3/(3), A1, staminodes usu petaloid, always modified into a labellum, lateral staminodes present/absent, G(3) inf, usu surmounted by epig glands, style often supported by a groove in the anther; ov n, ax/par. Fr usu capsule. *Trop.*

**239. Costaceae.** Lvs spirally arranged with closed sheaths, ± ligulate, not aromatic. Fls like *Zingiberaceae* but lacking external epig glands. *Trop.*

**240. Cannaceae.**† Lvs spirally arranged, eligulate. Infl terminal with fls in pairs. K3, C(3), A1 petaloid with ½-anther, + petaloid staminodes, G(3), inf, with petaloid style; ov n, ax. Warty capsule. *Trop America* (*Canna*).

**241. Marantaceae.**† Lvs distichous, petiole with ± swollen band at apex. Infl panicle/spike with asymmetrical fls in pairs. K3, C(3), A1 + various petaloid staminodes, G(3), inf; ov 1 per cell (2 sometimes aborting). Capsule, often fleshy. *Trop.*

## Superorder Arecidae

Inflorescence with numerous flowers subtended by a prominent spathe, often aggregated into a spadix; endosperm (when present) usu containing fats, oils, proteins and/or hemicelluloses; stomata usually with 2–4 subsidiary cells; leaves usually petiolate and broad, without typical parallel venation.

### ARECALES

Palms; usu with woody stems; lvs plicate, splitting.
**242. Palmae/Arecaceae.***† Trees/shrubs/prickly scramblers. Lvs large & plicate, becoming palmately/pinnately divided. Infl fleshy panicles/spikes, often with spathe-like bracts. Fls uni/bisex, act. P6/(6), fleshy, A6 or more, G usu (3), sup, 1–3-loc; ov 3; carpels rar free, 1-ovulate. Berry/drupe, sometimes huge. *Mainly Trop.*

### CYCLANTHALES

Usu palm-like herbs; lvs plicate, usu bifid; fls reduced.
**243. Cyclanthaceae.** Monoecious herbs/woody climbers. Lvs plicate, often deeply bifid. Infl a spadix subtended by caducous spathes. Male fls: P4/(4), A n; female: P cupular/0, staminodes 4, G1-loc, inf; ov n, par. Syncarp of berries. *Trop America.*

### PANDANALES

Woody plants with long stiff lvs; syncarps.
**244. Pandanaceae.** Dioecious trees/shrubs, often with stilt roots. Lvs crowded, leathery, keeled, often spinulose. Fls paniculate or in spadices, unisex. P rudimentary/0, A n, G1-loc, sup/naked; ov 1–n, basal/par. Fr syncarp, units woody/fleshy. *Old World Trop, Hawaii.*

### ARALES

Spathe & spadix usu present; lvs often petiolate & broad, not plicate.
**245. Araceae.***† Herbs/woody climbers, sap often bitter/milky, rar reduced floating aquatics. Lvs usu petiolate and broad, often lobed, venation often reticulate. Fls minute, sessile on spadix enclosed in conspicuous spathes, uni/bisex. P4–6/(4–6)/0, A2–8, G1–n-loc, sup/-naked; ov 1–n. Usu berry. *Trop (mainly) & Temp.*
**246. Lemnaceae.***† Small floating or submerged aquatics, thallus-like, without true roots. Fls unisex, P0, A1–2, G1-loc, naked; ov 1–7. Utricle. Often reproducing by budding. *Widespread.*

## Superorder Liliidae

Both perianth whorls usually petaloid; ovary syncarpous; pollen grains usually binucleate; endosperm, when present, seldom of starch; stomata usually without subsidiary cells.

LILIALES

P petaloid, usu act; A usu 6/3; G sup/inf; septal nectaries frequent.

**247. Pontederiaceae.**† Aquatic. Lvs with sheathing bases, often petiolate. Infl racemose in axil of spathe-like sheath. Fls bisex, act rar zyg. P(6), petaloid, A usu 6, G(3), sup; ov 3–n, ax/par. Capsule. *Trop & warm Temp.*

**248. Liliaceae** s.l.*† Habit diverse: herbs with rhizomes, corms, bulbs, etc/trees/shrubs/climbers. Lvs basal/cauline with main veins parallel to margin, sometimes succulent, spiny-margined, rar reduced to scales when cladodes present. Infl usu racemose/umbellate, if fls solit then hardly ever truly terminal on naked (ebracteate) scape. Fls bisex rar unisex, act/weakly zyg. P usu 6/(6), mostly petaloid, with/without corona, rar 2- or more than 3-merous, A6/(6) rar 4/more than 6/3 + 3 staminodes, G usu (3), sup/inf; ov n–3, ax/rar par. Capsule/berry. *Widespread.*

**249. Iridaceae.***† Terrestrial herbs with rhizomes/corms/bulbs. Lvs often equitant. Infl racemose/fls solit. Fls bisex, act/zyg. P6/(6) petaloid, whorls similar/dissimilar, A3, G(3) inf, with styles often divided; ov n–few, usu ax. Capsule. *Widespread.*

**250. Tecophilaeaceae.** Terrestrial herbs with corms/tubers. Lvs mostly basal. Infl raceme/panicle. Fls bisex, act. P6/(6) short-tubed, A6/3–4 + 2–3 staminodes, poricidal, G(3) partly inf; ov n, ax. Capsule. *Trop & S Africa, Chile.* (*Excl Cyanastraceae.*)

**251. Velloziaceae.** Dichotomously branched shrubs with persistent leaf bases/± woody-based herbs. Fls solit, terminal on naked peduncles in terminal tufts of lvs. P6/(6) petaloid, A6/more in 6 bundles, corona appendages sometimes present, G(3), inf; ov n, ax. Hard capsule, often spinulose/glandular. *Trop Arabia, Madagascar, Africa, S America.* (*Often placed close to Hypoxidaceae.*)

**252. Haemodoraceae.**† Terrestrial herbs, sap often orange. Lvs mostly basal, often equitant. Infl cymes/racemes/clusters. Fls bisex, act/weakly zyg. P6/(6) petaloid, persistent, often densely hairy, A6/3, G(3) sup/inf; ov n, ax. Capsule. *Mainly S Hemisphere; N America.*

**253. Taccaceae.** Scapose herbs. Lvs broad, often petiolate. Fls umbellate, involucrate, inner bracts of umbel often dangling. P(6) ± petaloid, A6, G(3) inf; ov n, par. Berry/capsule. *Trop & China.*

**254. Stemonaceae.** Erect/climbing herbs with tuberous roots. Lvs alt/-opp/whorled, cauline, petiolate with broad lamina. Fls axillary. P4 sepaloid/petaloid, A4, G1-loc, sup; ov n–2, apical/basal. Capsule 2-valved. *India to Japan & Australasia, SE USA.*
**255. Dioscoreaceae.***† Climbers with swollen rootstock. Lvs cauline, usu alt, petiolate, often cordate/digitate. Racemes axillary. Fls unisex, act, small. P6/(6), often greenish, A6/3, free/connate, G(3), sup; ov 2 per cell, ax. Capsule/berry. *Mainly Trop & warm Temp.*

ORCHIDALES

Fls zyg with 1 petal forming a labellum; A1–2; G ± inf; seeds with undifferentiated embryo.
**256. Orchidaceae.***† Terrestrial/epiphytic/saprophytic herbs. Lvs alt/rar opp. Fls in racemes/solit, bisex, zyg. K3 rar 2/(3), P3, both ± petaloid, median petal usu modified into labellum, A usu 1–2, adnate to stylar column, pollinia frequent, G(3), inf, usu twisted in flower (unless raceme pendulous); ov n, usu par. Capsule. *Widespread.*

# FURTHER IDENTIFICATION

The identification of the family to which a plant belongs is only the first necessary step in its complete identification. To make the key more generally useful we have provided below some notes on the more important literature which can be used for identification to generic and specific levels.

First, it must be stressed that the present key does not include all currently recognised flowering plant families. All exclusively tropical and southern hemisphere families have been excluded unless they contain plants fairly commonly cultivated in the northern hemisphere. Therefore, if a specimen cannot be satisfactorily identified to its family in this key, the following works, which are world-wide in scope, should be consulted:

G. BENTHAM & J. D. HOOKER, *Genera Plantarum*, 1862–83 (in Latin).

A. ENGLER & K. PRANTL, *Die Natürlichen Pflanzenfamilien*, 1887–99; 2nd edn, 1924 & proceeding (in German).

A. ENGLER (ed. H. MELCHIOR) *Syllabus der Pflanzenfamilien*, 12th edn, 1964 (in German).

J. HUTCHINSON, *The Families of Flowering Plants*, 2nd edn, 1959 (in English).

V. H. HEYWOOD (ed.) *Flowering Plants of the World*, 1978

A punched-card key to the families has been published by B. HANSEN & K. RAHN, *Dansk Botanik Arkiv* **26**(1), 1969.

The following books, though less wide in their scope (at least as far as identification is concerned) than those cited above, also contain much useful information:

A. B. RENDLE, *Classification of Flowering Plants*, vol. 1, 1930, vol. 2, 1938.

G. H. M. LAWRENCE, *Taxonomy of Vascular Plants*, 1951.

L. B. BENSON, *Plant Classification*, 1957.

Keys to the subfamilies or tribes of many large families (including *Compositae* and *Gramineae*) are given in Benson's book.

It is important to emphasise that the circumscription of a particular family can vary considerably from book to book. Care must be taken, therefore, to see that a family arrived at in the present key corresponds fairly closely with the family of the same name in another work in which further identification is to be made. This is often extremely difficult; it may, however, sometimes be done by checking the specimen against the family descriptions, synonymy and references given in all the works which are being used.

Having determined the family, one may proceed to the identification of genus and species. Particularly if the origin of the plant is unknown, the first step is usually to try to find out whether any world-wide monographic study of the family exists. The most notable series of such monographs is that begun by Engler under the general title *Das Pflanzenreich*, which covers many whole or part families. It is impossible for us to list these, but most large botanical libraries possess a full set. Many other revisions or monographs of families exist outside this series, as books, or as papers in numerous botanical journals. Lawrence's book (cited above) and also many modern Floras are very useful in this respect, as references to monographic works are given after the descriptions of each family included in them.

If a family monograph is not available, identification as far as the genus is possible with the following works:

BENTHAM & HOOKER (cited above), particularly valuable for its synopses of genera, which may enable a plant to be placed in its genus when the specimen is too incomplete to be run down in a key.

ENGLER & PRANTL (cited above).

HUTCHINSON (cited above) for most monocotyledonous and many of the smaller dicotyledonous families.

A. LEMÉE, *Dictionnaire descriptif et synonymique des Genres des Plantes phanérogames,* 1925–43. Vols 8a & 8b contain keys to all genera (in French).

J. HUTCHINSON, *The Genera of Flowering Plants*, vol. 1, 1964 & vol. 2, 1967 (the only two volumes completed).

C. D. K. COOK, *Water Plants of the World*, 1974. Includes a key to aquatic genera based on vegetative characters.

Monographs or revisions of genera are very numerous but there is no comprehensive list of them available. However, many herbaria and botanical institutions have lists or card-indexes of such works, and these should be consulted if possible.

If the specimen is of known geographical origin, identification to genus and species may conveniently be made if a Flora of the region is available. Lists of Floras are given by:

S. F. A. BLAKE & A. C. ATWOOD, *Geographical Guide to the Floras of the World: Part 1, Africa, Australasia, N & S America & Islands of the Atlantic, Pacific and Indian Oceans*, 1942; *Part 2, Western Europe*, 1961.

LAWRENCE (cited above), pp. 290–303.

C. L. PORTER, *Taxonomy of Vascular Plants*, 1959.

T. G. TUTIN, *et al.* (ed.) *Flora Europaea* (vols 1–5, 1964–79).

If the specimen is a cultivated plant, the following books should prove helpful:

C. K. SCHNEIDER, *Illustriertes Handbuch der Laubholzkunde*, 1904–12.

L. H. BAILEY, *Manual of Cultivated Plants*, 2nd edn, 1949.

A. REHDER, *Manual of Cultivated Trees and Shrubs hardy in North America*, 2nd edn, 1940.

ROYAL HORTICULTURAL SOCIETY (ed.) *Dictionary of Gardening* & Supplement, 1951–6.

G. KRÜSSMANN, *Handbuch der Laubgehölze*, 1962.

F. ENCKE (ed.) *Parey's Blumengärtnerei*, 2nd edn (1958).

The works cited above should assist in identification to the species; J. W. C. KIRK'S *A British Garden Flora* (1927) is helpful only to the generic level. When identifying garden plants it should be borne in mind that cultivated material may (for various reasons) differ somewhat from the wild stock.

It is often helpful to confirm an identification by comparison with an illustration; references to these will be found in *Index Londinensis* (ed. O. STAPF, 1921–41). Many popular, illustrated Floras (such as O. POLUNIN'S *Wild Flowers of Europe*, 1969) have been published in recent years; these are generally helpful, but it must be remembered that they offer only selective coverage of the plants of the areas they deal with. J. C. WILLIS'S *A Dictionary of Flowering Plants and Ferns* (6th edn, 1931, reprinted 1948, 7th edn, 1966 and 8th edn, 1975, both edited by H. K. AIRY SHAW) contains much useful information (e.g. for each genus, the citation of its family).

Finally, the value of comparing the identified specimen with named herbarium material cannot be overemphasised. This is the most stringent test of the accuracy of an identification, although the warning should be given that herbarium material is sometimes wrongly named. A herbarium may be used to by-pass the literature when dealing with small families – a rapid survey of the material of the family in question may be sufficient for identification. This is also possible with a small genus, but it is not to be recommended if the family or genus is large, or if its classification is known to be difficult. Study of herbarium material is also useful when the specimen to be named is too incomplete for identification by means of a key.

## TERMINOLOGY

Taxonomy is a subject bedevilled by a profuse terminology, much of it in, or derived from, Latin, as are the original descriptions of the groups, from family to species and below. W. T. STEARN's *Botanical Latin* (2nd edn, 1972) provides a comprehensive, well-illustrated and easily understood guide to the subject, as well as a detailed glossary. Other useful glossaries are:

LAWRENCE (cited above), pp. 737–75.

B. D. JACKSON, *A Glossary of Botanic Terms*, 4th edn, reprinted 1953.

H. I. FEATHERLY, *Taxonomic Terminology of the Higher Plants*, 1959, facsimile edition 1965.

Two very helpful polyglot botanical dictionaries are:

J. NIJDAM, *Woordenlijst voor de Tuinbouw in zeven Talen*, 1952 (Dutch, English, French, German, Danish, Swedish & Spanish).

N. N. DAVIDOV, *Botanicheskii Slovar'*, 1960 (Russian, English, German, French, Latin).

# GLOSSARY

Only very brief definitions are given here; if more detail is required, reference should be made to the glossaries cited on p. 100, or to a general botanical textbook.

*abaxial*: (of a lateral organ) the side away from the axis.
*achene*: a small, dry indehiscent 1-seeded fruit; in the strict sense, formed from 1 free carpel.
*actinomorphic*: regular, having 2 or more planes of symmetry (cf. p. 19).
*adaxial*: (of a lateral organ) the side towards the axis.
*adnate*: with one organ apparently fused to another (e.g. the stamens and corolla in *Salvia*).
*aestivation*: the way in which the perianth parts are arranged relative to one another in the bud (cf. p. 17).
*androecium*: the male sex organs (stamens) collectively.
*androgynophore*: the stalk bearing the ovary and stamens (e.g. *Passiflora*).
*anther*: that part of the stamen in which the pollen is produced.
*anthesis*: that period in the life history of a flower between the opening of the bud and the withering of the stigma or stamens.
*anthocyanins*: chemical compounds (pigments) found in most plants, but replaced by betalains in most Caryophyllales.
*antipetalous*: generally of stamens when they are the same number as, and opposite to (i.e. on the same radii as) the corolla segments (e.g. *Primula*).
*antisepalous*: (stamens) the same number as, and opposite to, the calyx segments.
*-aperturate*: (pollen grains) referring to the number of pollen tube germination sites (grooves or pores), e.g. 1-aperturate.
*apetalous*: lacking a corolla.
*apical*: (placentation) cf. p. 17.
*apocarpous*: having free carpels.
*aril*: an appendage borne on the seed, strictly an outgrowth of the funicle.
*asymmetric*: (corolla) having no planes of symmetry.
*axile*: (placentation) cf. p. 12.
*axillary*: in the axil (i.e. the junction between petiole and stem).

*basal*: (placentation) cf. p. 17.
*basifixed*: attached by the base.

*berry*: a fleshy, indehiscent fruit with the seeds immersed in pulp.
*betalains*: chemical compounds, characteristic of most Caryophyllales.
*bifid*: divided into 2 shallow segments, usually at the apex.
*bilabiate*: 2-lipped.
*binucleate*: (pollen grain) containing 2 nuclei when shed.
*bipinnate*: (leaf) a pinnately divided leaf with the segments themselves pinnately divided (e.g. *Acacia*).
*bract*: a frequently leaf-like organ (often very reduced or absent) bearing a flower, inflorescence or partial inflorescence in its axil.
*bracteole*: a frequently leaf-like organ (often very reduced or absent) borne on the pedicel.
*bulb*: an underground organ composed of stem and swollen leaf bases, enclosing the bud for next year's growth.

*caducous*: falling off early.
*calyptrate*: (usually of perianth) shed as a unit, often in the shape of a cap or candle-snuffer (e.g. calyx of *Eschscholtzia*).
*calyx*: the outer whorl(s) of the perianth, consisting of sepals.
*cambium*: meristematic tissue occurring initially between the xylem and the phloem.
*capitate*: (stigma) like a pin-head, or knob.
*capitulum*: a head of sessile or almost sessile flowers.
*capsule*: a dehiscent, usually dry fruit formed from a syncarpous ovary.
*carpel*: the organ containing the ovules; in syncarpous ovaries often much modified.
*caruncle*: an outgrowth near the hilum (the point of attachment) of a seed (adj. *carunculate*).
*caryopsis*: an achene with the seed adnate to the fruit wall.
*catkin*: a unisexual, elongate inflorescence of small, apetalous flowers, often deciduous as a whole.
*caudex*: the often woody zone joining root and stem.
*cauline*: borne on the stem.
*centrifugal*: (stamens) developing serially from the inside outwards.
*centripetal*: (stamens) the opposite of centrifugal.
*cincinnus*: a short, coiled cyme.
*cladode*: a lateral, usually flattened stem structure borne in the axil of a reduced leaf.
*claw*: the conspicuously narrowed and attenuate base of an organ, especially a petal (adj. *clawed*).
*compound*: (leaf) divided into distinct and separate leaflets.
*connate*: with the parts of the same whorl fused to each other (e.g. petals in gamopetalous flowers).
*connective*: the part of the stamen joining the anther cells.
*contorted*: (aestivation) cf. p. 17.
*cordate*: heart-shaped.

*corona*: an outgrowth (frequently petaloid) of the corolla, stamens or staminodes.
*corymb*: a flat-topped inflorescence in which the branches arise at different levels.
*cotyledon*: the first seedling leaf.
*cupule*: a cup formed from free or united 'bracts', often containing an ovary.
*cyme*: a determinate or centrifugal inflorescence (adj. *cymose*).
*cypsela*: a small, indehiscent, dry, 1-seeded fruit, formed from an inferior ovary; (e.g. *Compositae*); often loosely termed 'achene'.
*cystolith*: a mineral concretion found in special cells in certain leaves (e.g. *Urticaceae*).

*deciduous*: (leaves) falling once in each year; also used of stipules, catkins, sepals.
*dehiscence*: the mode of opening of an organ, usually anther or fruit (adj. *dehiscent*).
*determinate*: (inflorescence) one in which the terminal flower opens first and stops further growth of the axis.
*dichasium*: a cyme in which the opposite branches are more or less equal, with a flower or another branch in the fork between them (adj. *dichasial*).
*diffuse-parietal*: (placentation) cf. p. 14.
*digitate*: palmate into many narrow segments (e.g. leaf of *Lupinus*).
*dimorphic*: of two forms.
*dioecious*: with male and female flowers on separate plants.
*disc*: a fleshy, nectariferous organ (circular or lobed) frequently developed between the stamens and ovary, sometimes outside the stamens also.
*disjunct*: (distribution) found in two or more separate areas.
*distichous*: (leaves, flowers borne in spikelets) alternating in two opposite ranks.
*divided*: (leaves) cf. p. 18.
*drupe*: a fleshy or leathery 1-few-seeded fruit with a hard inner wall.

*endosperm*: storage material in many seeds, formed after fertilisation.
*entire*: (leaves) simple and with smooth margins.
*epicalyx*: a whorl of sepaloid segments borne outside the true calyx (e.g. *Potentilla*, *Malva*).
*epidermis*: the outermost cell-layer of plants.
*epigynous*: cf. pp. 4–12.
*epipetalous*: borne on the corolla.
*epiphytic*: growing on another plant, but not parasitic.
*equitant*: (leaves) folded sharply inwards from the midrib, the outermost leaf enclosing the next, etc. (e.g. *Iris*).
*ericoid*: looking like *Erica*, particularly the leaves.
*exarillate*: without an aril.
*exfoliating*: (bark) scaling off in large flakes.
*exotic*: native only from outside a specified area (i.e. in the case of this book, native only from south of 30° N).
*exstipulate*: without stipules.
*extrorse*: (anthers) opening towards the outside of the flower.

*false septum*: a secondary septum in an ovary, formed after the formation of the primary septa (e.g. *Linum*); also used in reference to the fruits of *Cruciferae*.
*fascicle*: a small bunch of pedicillate flowers, or a bundle of stamens (adj. *fascicular*, *fasciculate*).
*filament*: the stalk of a stamen, bearing the anther.
*filiform*: thread-like.
*foliaceous*: leaf-like.
*-foliolate*: divided into a specified number of separate leaflets (e.g. 3-foliolate).
*follicle*: a 1-carpellate, several-seeded fruit (or partial fruit), dehiscing along one suture, usually the adaxial (e.g. *Delphinium*).
*free-central*: (placentation) cf. p. 14.
*fruit*: the structure(s) containing all the seeds produced by a single flower.
*funicle*: the stalk connecting an ovule to its placenta.

*gamopetalous*: with united corolla segments.
*geniculate*: (filaments) jointed, knee-like.
*glabrous*: without hairs.
*gland*: a secretory organ.
*globose*: more or less spherical.
*gynobasic*: (style) attached near the base of the carpels.
*gynoecium*: the female sex organs collectively.
*gynophore*: the stalk of a stalked ovary.

*halophytic*: growing on saline soils.
*hemiparasitic*: partial parasites with green leaves.
*herb*: applied to plants dying back more or less to ground level during each unfavourable season; including plants woody only at the base, and annuals.
*heteromorphic*: having two or more forms.
*hypogynous*: cf. pp. 4–12.

*imbricate*: (aestivation) cf. p. 17.
*indehiscent*: (fruit) not opening by any definite mechanism.
*indeterminate*: (inflorescence) one in which the lower or outer flowers open first, and the axis continues growth.
*indusiate*: borne within a sheath.
*inferior*: (ovary) cf. p. 5.
*internode*: that part of a stem between one leaf base and the next.
*interpetiolar*: (stipules) situated between the bases of opposite leaves.
*intrapetiolar*: (stipules) situated between the stem and the upper surface of the petiole.
*intraxylary*: (phloem) nearer to the centre of the stem than the xylem is.
*introrse*: (anthers) opening towards the inside of the flower.
*intrusive*: (placenta) cf. p. 14.
*involucel*: a cup-like structure surrounding the inferior ovary of *Dipsacaceae*.
*involucre*: a series of bracts (often overlapping) surrounding an inflorescence (adj. *involucrate*).

*labellum*: a lip; generally used of the enlarged abaxial petal of some monocotyledonous flowers.
*laciniate*: deeply slashed into narrow divisions.
*leaflet*: the separate parts into which compound leaves are divided; distinguished from leaves by their not having buds in their axils.
*legume*: a 1-carpellate, dry, dehiscent fruit, usually several-seeded, dehiscent along both sutures; characteristic of *Leguminosae.*
*lepidote*: bearing peltate, often scurfy scales; or such scales themselves.
*ligule*: a tongue-like outgrowth on a petal (e.g. *Silene*), or at the junction of leaf-sheath and blade (e.g. *Gramineae*); adj. *ligular* or *ligulate*.
*linear*: at least 12 times longer than broad, with the sides more or less parallel.
*loculicidal*: dehiscence of a capsule along the abaxial line(s) of the carpels, i.e. down the middles of the loculi.
*loculus*: the cavity(ies) in a carpel, ovary or anther.
*lodicules*: small, often scale-like organs borne below the stamens in grass flowers.
*lomentum*: an indehiscent fruit which fragments transversely between the seeds.

*marginal*: (placentation) cf. p. 12.
*medifixed*: (hairs, anthers) attached by the middle.
*membranous*: thin and translucent, not hardened.
*mericarp*: a 1-seeded portion of an initially syncarpous fruit which splits apart at maturity (e.g. most *Umbelliferae*).
*-merous*: indicating number of parts (e.g. 3-merous or trimerous).
*monadelphous*: (stamens) with the filaments united into a tube.
*monoecious*: with male and female flowers on the same plant.
*multifid*: deeply divided into numerous lobes.

*naked*: (ovary) cf. p. 8.
*nut*: a hard, indehiscent, 1-seeded fruit.
*nutlet*: a dry, 1-seeded nut-like, partial mericarp, characteristic of *Labiatae* and *Boraginaceae*.

*obconical*: in the form of an inverted cone.
*obdiplostemony*: having the stamens twice as many as the petals, the outer whorl of stamens opposite the petals (e.g. *Geranium*).
*obsolete*: very reduced, almost absent.
*ovules*: the structures in the ovary which become seeds after fertilisation.

*palmate*: (leaves) divided to the base into separate leaflets, all the leaflets arising from the apex of the petiole.
*panicle*: a much-branched inflorescence.
*papilionaceous*: with a flower like that of *Pisum*.
*pappus*: a ring of hairs or scales on top of a cypsela.
*parietal*: (placentation) cf. p. 14.
*partly inferior*: (ovary) cf. p. 5.

*pedicel*: the stalk of a single flower in an inflorescence (adj. *pedicillate*).
*peduncle*: the stalk bearing an inflorescence or solitary flower.
*peltate*: disc-shaped, the stalk arising from the under-surface.
*perianth*: the outer, sterile whorls of a flower, often differentiated into calyx and corolla.
*pericarp*: the fruit wall.
*perigynous*: cf. pp. 4–12.
*perisperm*: storage tissue in some seeds, formed from the nucellus (i.e. maternal tissue).
*petal*: a single segment of the corolla.
*petaloid*: having the colour and texture of a petal.
*petiole*: the 'stalk' of a leaf, attaching the leaf blade to stem or branch.
*phloem*: the internal tissue involved in the conduction of organic solutes.
*phyllodic*: (leaves) having flattened, leaf-like petioles instead of a true lamina.
*phyllotaxis*: the manner in which the leaves are arranged: generally expressed as a fraction – the number of complete turns made around the axis, counting from leaf to leaf until a leaf immediately above the original is reached, over the number of leaves passed in reaching this leaf.
*pinnate*: (leaf) bearing separate leaflets along each side of a common stalk.
*pinnatisect*: pinnately divided nearly to the midrib.
*pistillode*: a rudimentary, non-functional gynoecium.
*placenta*: that part of the carpel wall or axis on which the ovules are borne.
*plicate*: folded in many pleats.
*pollinia*: pollen-bodies formed from all the pollen in the anther loculi, characteristic of many *Asclepiadaceae* and *Orchidaceae*.
*polygamous*: an inflorescence containing both unisexual and bisexual flowers.
*polymerous*: with numerous parts.
*polypetalous*: with distinct, free petals.
*pome*: a fleshy 'fruit' in which the carpels are immersed in the flesh (e.g. apple).
*poricidal*: (anthers) opening by apparently terminal pores.
*punctate*: dot-like, often used for glands or stigmas.

*raceme*: a simple, elongate, indeterminate inflorescence with pedicillate flowers (adj. *racemose*).
*radical*: (leaves) arising directly from the rootstock.
*radicle*: the part of the embryo which grows into the root.
*receptacle*: the enlarged top of a peduncle bearing crowded flowers (e.g. *Compositae*).
*rhizomatous*: having rhizomes, i.e. underground stems bearing scale leaves and adventitious roots.
*rosulate*: (leaves) in rosettes.

*saccate*: (perianth or corolla) with a conspicuous, hollow swelling.
*sagittate*: arrow-head-shaped.

*samara*: a dry, winged, indehiscent fruit or mericarp, usually 1-seeded (e.g. *Fraxinus*).

*saprophyte*: a plant which obtains its food materials by absorption of complex organic chemicals from the soil; often without chlorophyll.

*scape*: a (leafless) peduncle arising directly from a rosette of basal leaves, as in *Taraxacum* (adj. *scapose*).

*scarious*: 'glassy' and hardened.

*schizocarp*: a syncarpous fruit which splits into separate mericarps.

*scorpioid*: (cyme) a coiled, determinate inflorescence (e.g. *Myosotis*).

*sepal*: a single segment of the calyx.

*septicidal*: dehiscence of a capsule through the septa or carpel margins.

*septum*: the partitions in an ovary or fruit, formed from the united carpel margins (adj. *septate* (of ovaries) or *septal* (of nectaries borne on the septa)).

*-seriate*: in a number of series, e.g. 1-seriate or uniseriate.

*serrate*: with sharp, more or less regular teeth, like a saw.

*sessile*: not stalked.

*setaceous*: thread- or bristle-like.

*simple*: (leaves) not divided into separate leaflets.

*sinus*: the gap between the bases of two projecting organs.

*spadix*: a fleshy spike of numerous small flowers (e.g. *Arum*).

*spathe*: a large bract sheathing an inflorescence (often a spadix, as in *Arum*).

*spike*: a raceme in which the flowers (or in grasses, spikelets) are sessile on the axis (adj. *spicate*).

*spikelet*: a secondary spike, a group of 1 or more flowers subtended by bracts, as in *Gramineae* and *Cyperaceae*.

*spinulose*: with small spines.

*spur*: (of perianth or corolla) a long, usually nectariferous, tubular projection (e.g. *Viola*).

*stamen*: the male sex organ of a flower; usually consisting of anther, connective and filament.

*staminode*: a sterile stamen (adj. *staminodial*).

*stellate*: (hair) compound and star-shaped.

*stigma*: the receptive part(s) of the gynoecium, on which the pollen germinates.

*stipulate*: bearing stipules.

*stipules*: a pair of lateral outgrowths arising at the base of the petiole (e.g. *Trifolium*).

*stipuloid*: resembling stipules.

*style*: the often elongated portion of the ovary, bearing the stigma(s) at its apex or along its side.

*superior*: (ovary) cf. p. 4.

*superposed*: (ovules) in a single vertical row in each loculus; (spikes) one above the other, each separately stalked.

*supra-axillary*: (branches, inflorescences) borne above the axil from which they originate.

*sympetalous*: gamopetalous.

*syncarp*: a multiple or aggregate fruit, often fleshy or woody (e.g. *Morus*, *Pandanus*).
*syncarpous*: with united carpels.
*syngenesious*: (stamens) with the anthers united into a tube, the filaments free.

*tendril*: a sensitive, thread-like organ, coiling around objects touched.
*ternate*: a division of a leaf or leaflet into 3 parts.
*tetrads*: groups of 4 pollen grains which are shed as units.
*thallus*: a plant body undifferentiated into stem and leaf (adj. *thalloid*).
*torus*: the apex of the pedicel, bearing the floral appendages.
*translator*: a specialised structure uniting the pollinia (in most *Asclepiadaceae* and *Orchidaceae*).
*trifid*: shortly divided into 3.
*trifoliolate*: (leaves) divided into three leaflets.
*trinucleate*: (pollen) containing three nuclei when shed.
*triquetrous*: three-sided.
*truncate*: ending abruptly, as though broken off.

*umbel*: a usually flat-topped inflorescence in which all the pedicels arise from the same point on the peduncle (e.g. most *Umbelliferae*).
*utricle*: a bladdery, indehiscent, 1-seeded fruit.

*valvate*: (aestivation) cf. p. 17.
*valvular*: (anthers) opening by valves or small flaps (e.g. *Berberis*).
*versatile*: (anthers) pivoting freely on the filament.
*verticillate*: (inflorescence) the flowers in superimposed whorls, each whorl consisting of two opposite (often modified) cymes (e.g. *Mentha*).
*vessels*: water-conducting elements of the xylem, characterised by lignified side walls, cross-walls obsolete or absent.
*villous*: clothed in long, woolly hairs.

*xeromorphic*: with the habit of plants characteristic of arid regions – e.g. with reduced or fleshy leaves, or densely hairy, etc.
*xylem*: the woody tissue of plants, involved in the upward conduction of water.

*zygomorphic*: bilateral, having one plane of symmetry only.

# INDEX TO FAMILIES AND HIGHER GROUPS

The index covers names of families and higher groups included in the keys and descriptions, as well as some synonyms. Accepted names are in Roman type, others in *italic*.